Foreword

The *Specification for the construction of slurry trench cut-off walls* has been drawn up for use where slurry trench cut-off walls are required to act as barriers to pollution migration. It should be read in conjunction with the accompanying Notes for Guidance. These documents aim to provide a standard and consistent approach to the design, construction, testing and monitoring of cut-off walls and guidance on the appropriateness of this technique and current best practice.

The Specification and Notes for Guidance relate to the use of slurry trench cut-off walls for the control of migration of groundwater, leachate, chemical contaminants and gases. Such cut-off walls, which often include low permeability geomembranes, are excavated under the support of a self-hardening cementitious slurry, i.e. as a single-phase process. Cut-off walls constructed by driven methods, two phase processes, or using soil-bentonite mixes, are not covered by this Specification. Further guidance on such methods is contained in the bibliography in the Notes for Guidance.

The Specification makes provision for a Particular Specification to incorporate project-specific requirements, information and other matters.

This Specification is intended to be compatible with use of the *ICE Conditions of Contract* (7th edition) and the *Civil engineering standard method of measurement* (3rd Edition). Where the Specification is to be used with other contractual documents the user may need to make amendments as appropriate.

Acknowledgements

The Specification and these Notes for Guidance are presented as an authoritative and unifying statement of good practice. They comprise one discrete set in a series of specifications commissioned by the Ground Board of the Institution of Civil Engineers, in this case in conjunction with the Building Research Establishment and the Construction Industry Research and Information Association. The documents were prepared by Mr Graham Doe of Amec Piling and Foundations, Dr Stephan Jefferis of Golder Associates and Dr Paul Tedd of the Building Research Establishment, under the guidance of and with the assistance of a small Working Party of engineers expert and experienced in the area. Drafts of the documents were reviewed by a wide body of practising engineers.

The Working Party comprised:

Mrs Sue Herbert (Chairman)	The Environment Agency (formerly Stanger Science & Environment) and representing the ICE Ground Board
Mr Fin Jardine	CIRIA (replaced Dr Simon Johnson in January 1995)
Dr Paul Tedd	Building Research Establishment
Mr Peter Chamley	Ove Arup and Partners
Dr David Greenwood	Formerly Cementation Piling and Foundations

The Institution is indebted to the authors, the Working Party and to those who assisted them for their considerable efforts in preparing these documents. Mr Russell Jones of Golder Associates who provided advice on the properties and testing of geomembrane materials, Mr P J Barker of Bachy Soletanche, and Mr R D Essler of Keller Ground Engineering, provided valuable comment on the drafts.

The work was sponsored, in part, by the Department of the Environment, Transport and the Regions under their Technology and Performance Business Plan.

Contents

S pages refer to the Specification, N pages refer to the Notes that follow it.

		S	N
Introduction — General issues			1
1.	**General requirements for slurry trench cut-off walls**	1	5
1.1.	Standards and definitions	1	5
	1.1.1. British Standard Specifications	1	
	1.1.2. Codes of Practice	1	
	1.1.3. Definitions	1	
1.2.	General contract requirements	1	5
	1.2.1. Precedence of specification	1	
	1.2.2. Contractor to work to other Contractor's drawings	1	
	1.2.3. Progress report	1	
1.3.	Performance	2	
	1.3.1. Responsibility for execution and performance	2	5
	1.3.2. Method statement	2	
1.4.	Description of the slurry trench works	2	
1.5.	Particular specification	2	6
1.6.	Materials	3	6
1.7.	Site conditions	3	6
	1.7.1. Topographical information	3	6
	1.7.2. Geotechnical information	3	6
	1.7.3. Chemical information	3	7
1.8.	Workmanship	3	7
	1.8.1. Suitability of equipment	3	
	1.8.2. Reinstatement	3	7
1.9.	Access	4	7
1.10.	Working areas	4	8
1.11.	Disposal of waste arisings	4	8
1.12.	Setting out	4	8
	1.12.1. Responsibility	4	
	1.12.2. Setting out points	4	
1.13.	Programme	4	
2.	**Wall requirements**	5	9
2.1.	Toe depth	5	9
2.2.	Capping requirements	5	10

	2.3.	**Fluid properties of slurry**	5	11
		2.3.1. Slurry stability, bleeding	5	11
		2.3.2. Slurry loss to ground, penetration and filtration	5	12
	2.4.	**Set properties of slurry**	5	13
		2.4.1. Permeability	5	13
		2.4.2. Strength	5	13
		2.4.3. Physical and chemical durability	5	15
		2.4.4. Pre-construction mix trials	5	17
		2.4.5. Timing	5	17
	2.5.	**Geomembranes**	6	18
3.	**Materials**		7	19
	3.1.	**Slurry components**	7	19
		3.1.1. Water	7	19
		3.1.2. Bentonite powder	7	19
		3.1.3. Portland cement	7	20
		3.1.4. Ground granulated blastfurnace slag	7	20
		3.1.5. Pulverised fuel ash	7	20
		3.1.6. Certification	7	20
		3.1.7. Admixtures	7	20
	3.2.	**Geomembrane**	7	21
		3.2.1. Geomembrane material	7	22
		3.2.2. Manufacturer's quality control	8	24
		3.2.3. Delivery, handling and storage	8	24
		3.2.4. Contractor's experience/manufacturer's experience	8	24
		3.2.5. Welded joints	8	24
		3.2.6. Interlocking joints	9	24
4.	**Slurry production and quality control**		10	25
	4.1.	**Slurry preparation plant**	10	25
	4.2.	**Batching**	10	25
		4.2.1. Accuracy	10	25
		4.2.2. Calibration	10	25
	4.3.	**Quality control**	10	25
		4.3.1. Bentonite suspension	10	26
		4.3.1.1. Sampling	10	26
		4.3.1.2. Viscosity	10	26
		4.3.1.3. Specific gravity	10	26
		4.3.2. Cementitious slurry	11	26
		4.3.2.1. Sampling	11	26
		4.3.2.2. Specific gravity	11	27
		4.3.2.3. Bleed	11	27
	4.4.	**Temperature**	11	27
5.	**Slurry wall construction**		12	28
	5.1.	**Slurry wall construction**	12	28
		5.1.1. Excavation	12	28
		5.1.1.1. Excavation depth	12	28
		5.1.1.2. Guide walls	12	28

		5.1.1.3.	Slurry level		12	29
		5.1.1.4.	Ground and toe levels		12	29
	5.1.2.	Wall location and dimensions			12	29
		5.1.2.1.	Setting out		12	29
		5.1.2.2.	Width		12	29
		5.1.2.3.	Verticality		12	29
		5.1.2.4.	Toe levels, depths		12	30
	5.1.3.	Day joints			13	30
	5.1.4.	Temporary protection			13	30
		5.1.4.1.	Safety		13	30
		5.1.4.2.	Reduction of surface cracking		13	30
	5.1.5.	Actions on loss of slurry or collapse of trench			13	30
	5.1.6.	Excessive specific gravity of slurry in trench			13	30
	5.1.7.	Ambient temperature			13	31
	5.1.8.	Capping			13	32
		5.1.8.1.	Trimming of top of slurry		14	32
		5.1.8.2.	Capping detail		14	32
5.2.	**Additional measures for geomembrane walls**				**14**	**32**
	5.2.1.	Installation of geomembrane			14	32
	5.2.2.	Dimensions			14	32
	5.2.3.	Base levels, bottom details and sloping strata			14	32
	5.2.4.	Positional tolerance			14	33
	5.2.5.	Changes in ground level			14	33
	5.2.6.	Repairs to damaged sections and joints			14	33
	5.2.7.	Temporary stop ends			15	33
	5.2.8.	Temporary fixing and protection of geomembrane			15	33
	5.2.9.	Contractor's supervision			15	
	5.2.10.	Good practice			15	
6.	**Compliance testing for material properties**				**16**	**34**
6.1.	**Sampling for testing for set properties**				**16**	**34**
	6.1.1.	Sample tubes			16	34
	6.1.2.	Frequency			16	35
	6.1.3.	Location			16	35
	6.1.4.	Care of samples			16	35
	6.1.5.	Storage and transportation			16	35
6.2.	**Laboratory testing**				**16**	**36**
	6.2.1.	Approved laboratory			16	36
	6.2.2.	Scheduling of tests			16	36
	6.2.3.	Preparation and care of samples			17	36
	6.2.4.	Tests			17	36
		6.2.4.1.	Triaxial permeability tests		17	36
		6.2.4.2.	Unconfined compression strength tests		17	37
		6.2.4.3.	Reporting of results		17	
6.3.	**In-situ testing**				**17**	**37**
6.4.	**Geomembrane**				**17**	**38**
	6.4.1.	Geomembrane conformance testing			17	38
	6.4.2.	Geomembrane non-destructive testing			18	38

	6.4.3.	Geomembrane qualitative destructive testing	18	38
	6.4.4.	Geomembrane quantitative destructive testing	18	38
	6.4.5.	Geomembrane interlock performance	18	

7. Records — 19, 39

    7.1. Slurry wall records — 19

    7.2. Additional records for geomembrane walls — 19
        7.2.1. 'As-installed' plan — 19
        7.2.2. Geomembrane fault plan — 19

8. Post-construction monitoring — 20, 40

Appendix A. Tests for bentonite and slurry fluid properties — 21

    A.1. Marsh funnel viscosity test — 21
        A.1.1. Equipment — 21
        A.1.2. Procedure — 21

    A.2. Direct reading viscometer — 21
        A.2.1. Equipment — 21
        A.2.2. Procedure — 22
        A.2.3. Calculation — 22

    A.3. Specific gravity test — 22
        A.3.1. Equipment — 22
        A.3.2. Procedure — 23

    A.4. Bleed test — 23
        A.4.1. Equipment — 23
        A.4.2. Procedure — 23

Appendix B. Tests for set slurry properties — 24

    B.1. Determination of permeability in triaxial cell — 24

    B.2. Determination of the unconfined compressive strength — 24

Appendix C. Typical example report forms — 25

Bibliography — 41

1. General requirements for slurry trench cut-off walls

1.1. Standards and definitions

1.1.1. British Standard Specifications

All materials and workmanship shall be in accordance with the appropriate British Standards current at the time of tender, including those listed in this Specification, except that where the requirements of British Standards are in conflict with this Specification, the latter shall take precedence.

1.1.2. Codes of Practice

All work shall be carried out generally in accordance with the principles of relevant codes of practice current at the time of tender, including those referred to in this Specification.

1.1.3. Definitions

In this Specification the terms 'approved', 'approval' and 'required' mean 'approved by the Engineer', 'approval of the Engineer' and 'required by the Engineer' respectively. In addition, the following definitions shall apply:

'Slurry trench works':	the slurry trench cut-off wall including any geomembrane and wall capping.
'Bentonite suspension' or 'mud':	a suspension consisting of natural sodium bentonite or sodium-activated bentonite powder mixed with water.
'Cementitious slurry' or 'slurry':	a mixture of bentonite suspension, ordinary Portland cement, and may include either pulverised fuel ash (pfa) and/or ground granulated blastfurnace slag (ggbs) with or without admixtures.
'HDPE':	high density polyethylene.

1.2. General contract requirements

1.2.1. Precedence of specification

All materials and work shall be in accordance with this Specification. Where there may be a conflict in requirements between the Specification and the Particular Specification for any aspect of the slurry trench cut-off wall requirements, the Particular Specification shall take precedence.

1.2.2. Contractor to work to other Contractor's drawings

The Contractor, where so directed by the Engineer, shall be required to work to other Contractor's drawings whenever drawings for temporary works not included in the Contract are related to particular details of the slurry trench works.

1.2.3. Progress report

The Contractor shall submit to the Engineer on the first day of each week, or at such longer periods as the Engineer may from time to time direct, a progress report showing the current rate of progress and progress during the previous period on significant items of each section of the slurry trench works.

1.3. Performance

1.3.1. Responsibility for execution and performance

The slurry trench works shall be constructed by a specialist Contractor with relevant previous experience, who shall be responsible for the execution of the slurry trench cut-off wall. The Particular Specification will state who is responsible for the performance of the slurry trench cut-off wall in accordance with the specified requirements; this will depend upon whether the Particular Specification is in a prescriptive or performance format. The Particular Specification will be either prescriptive or performance in format but not a mixture of both.

A prescriptive Particular Specification shall provide complete details of the slurry cut-off wall mix and/or geomembrane specification and type to be used, together with full details of the approved suppliers.

A performance style of Particular Specification shall state the performance criteria of the hardened slurry or the slurry wall system and by which technique(s) they are to be measured. It will also allow the Contractor unrestricted choice of materials subject to approved quality assurance/quality control measures.

1.3.2. Method statement

The Contractor shall submit with his tender a statement outlining his proposed method of working and details of the plant to be used. This statement shall include but not be limited to:

(i) method of excavation
(ii) where a geomembrane is to be included, details of the proposed geomembrane and interlock
(iii) programme of the slurry trench works.

1.4. Description of the slurry trench works

The slurry trench works shall comprise the construction of a cut-off barrier to the detail and dimensions specified using self-hardening cementitious slurry in accordance with the Conditions of Contract, Drawings and Bills of Quantities. Where a geomembrane is required it shall be HDPE unless it can be shown to the satisfaction of the Engineer that an alternative material is in no way inferior to HDPE or such other material as specified in the Particular Specification.

1.5. Particular Specification

The following matters are described in the Particular Specification together with other project specific matters:

(a) The purpose of the cut-off wall including overall and detailed design philosophy and the media to be contained
(b) The wall dimensions (width and depth/founding level)
(c) Whether a geomembrane is to be included, its depth of toe into a defined stratum, physical dimensions, and the geombrane material if not HDPE
(d) Either full details of the slurry materials and mix proportions and, where appropriate, the geomembrane specification and type with details of approved suppliers; or the performance criteria of the hardened slurry, the geomembrane and the techniques by which they are to be measured, together with any quality assurance/ quality control measure requirements
(e) The testing regime required and parameters to be achieved to address the issues in Clause 2.3.2
(f) The testing regime required and parameters to be achieved to address the issues in Clause 2.4.3
(g) The period required for any preliminary on-site mix trials and excavation trials
(h) Testing requirements for demonstrating the chemical or physical or other compatibility of the slurry wall system

including any geomembrane with the site conditions, together with time allocation for this testing programme, which shall be carried out in advance of commencement of the slurry trench works.

    (i) Time periods allocated for:

        (1) carrying out any trials for mix design and durability
        (2) the required ageing period prior to sample testing
        (3) laboratory testing, reporting and subsequent analysis periods
        (4) durability – physical, chemical testing regime.

(j) The permitted working hours, and noise levels
(k) The number of sample tubes to make one representative sample set of slurry
(l) Any permeability requirements and testing regime of the geomembrane interlock
(m) Any required upper limit on panel length
(n) The slurry specific gravity upper limit
(o) Any in-situ testing requirements of the slurry wall system.

1.6. Materials

At least fourteen days prior to mobilisation to site for commencement of the slurry trench works, the Contractor shall provide for the Engineer's approval details of the proposed suppliers of materials including water and, where possible, the sources. Once approval has been given, the material and water suppliers shall not be changed without further approval by the Engineer in writing.

1.7. Site conditions
1.7.1. Topographical information

Prior to tender, factual information including drawing(s) which clearly show surface and underground natural and artificial features will be made available to the Contractor.

1.7.2. Geotechnical information

Prior to tender, factual information and interpretative reports on the site investigations for the slurry trench works will be made available to the Contractor. This information will include but not be limited to:

(a) soil and rock stratification
(b) soil and rock properties
(c) groundwater levels and movements.

1.7.3. Chemical information

Prior to tender, factual information on the chemical conditions at the site will be made available to the Contractor for the purposes of health and safety assessments, which shall include but not be limited to:

(a) ground chemistry
(b) groundwater chemistry
(c) details of gases present and the relevant concentrations and information on the in-situ pressures and flow rates.

1.8. Workmanship
1.8.1. Suitability of equipment

The Contractor shall satisfy the Engineer regarding the suitability, efficiency and adequacy of the equipment to be employed. The Contractor shall state the type and number of equipment to be used.

1.8.2. Reinstatement

On completion of the construction of the cut-off wall the working area shall be levelled and reinstated to the original condition prior to

1.9. Access Access routes to, from, and about the site shall be only as shown on the Drawings, unless prior approval of alternatives is obtained from the Engineer.

1.10. Working areas The limits of areas of site to be used during construction of the cut-off wall shall be as shown on the Drawings, unless prior approval of alternatives is obtained from the Engineer.

1.11. Disposal of waste arisings The Contractor shall furnish documentation to the Engineer as evidence of compliance with current legislation.

1.12. Setting out
1.12.1. Responsibility Before construction of the cut-off wall begins, the Contractor shall mark out the position of the wall with suitable identifiable pins or markers.

1.12.2. Setting out points The Contractor shall provide and maintain co-ordinated setting out and level points throughout the duration of the slurry trench works.

1.13. Programme The Contractor shall inform the Engineer at regular mutually agreed intervals of the forward programme for the slurry trench works.

2. Wall requirements

2.1. Toe depth

The cut-off wall shall be constructed to the depth shown in the Particular Specification and Drawings or to such depth as instructed by the Engineer or his representative on site following his monitoring of the excavation at intervals of chainage not exceeding 2.5 m apart or such other interval as shown in the Particular Specification. The Engineer's representative on site shall check the wall depth with the Contractor at these intervals so that he can sign the Daily Slurry Trench Report as a record of the slurry trench works.

2.2. Capping requirements

As soon as practicable following installation of the slurry wall it shall be trimmed down to the specified level and the trimmed surface covered with a protective capping of suitable cohesive material placed at a minimum compacted shear strength of 50 kPa to the dimensions shown on the Drawings.

2.3. Fluid properties of slurry
2.3.1. Slurry stability, bleeding

The slurry supplied to the trench shall be in such condition that the mix components do not segregate and that the bleed is no higher than that permitted in Clause 4.3.2. Measures shall be taken to correct any drop in solids level in the trench, caused by bleed water.

2.3.2. Slurry loss to ground, penetration and filtration

The mix design of the slurry shall take into account the particle size distribution shown in the site investigation report, and the effects that this may have on slurry loss to the ground by penetration and filtration together with any associated trench stability problems.

2.4. Set properties of slurry

The following properties are to be demonstrated throughout the course of the slurry trench works on samples taken at varying depths from the slurry in the trench while still fluid.

2.4.1. Permeability

A target permeability of less than 1×10^{-9} m/s is required. However, due to inherent variability of trench mixes, sampling and testing, at least 80% of results shall be less than 1×10^{-9} m/s and at least 95% of the results shall be less than 1×10^{-8} m/s, with no individual result in excess of 5×10^{-8} m/s, when measured in apparatus and tested as described in Appendix B at an age of 90 days or such later age as specified in the Particular Specification.

2.4.2. Strength

The minimum unconfined compressive strength at an age of 28 days shall be 100 kPa when measured in apparatus and tested as described in Appendix B.

2.4.3. Physical and chemical durability

Testing for the indication of attainment of adequate durability characteristics shall be carried out as specified in the Particular Specification.

2.4.4. Pre-construction mix trials

All pre-construction mix trials required by the Particular Specification shall be carried out in an identical manner in all respects using the same materials, mixer(s) and water sources as proposed for the slurry trench works.

2.4.5. Timing

The following shall be undertaken in the time periods allocated in the Particular Specification:

(a) carrying out any trials for mix design and durability
(b) the required ageing period prior to sample testing
(c) laboratory testing, reporting and subsequent analysis periods
(d) durability—physical, chemical testing regime.

2.5. Geomembranes Requirements for the use or otherwise of a geomembrane, its depth and physical dimensions, are stated in the Particular Specification. Where this so indicates, the requirements of Clause 3.2 shall apply.

3. Materials

All materials stored on site shall be kept in conditions which prevent contamination and deterioration.

3.1. Slurry components

3.1.1. Water

Water shall be potable, comply with BS 3148 and be supplied from the local mains. Where this is not possible, alternative sources will be approved by the Engineer, provided satisfactory compatibility can be demonstrated by the Contractor by testing prior to commencement of the slurry trench works.

3.1.2. Bentonite powder

Bentonite powder shall be natural sodium bentonite or sodium-activated bentonite to CE grade or other bentonite that has been proven successful in trials or previous slurry wall construction.

3.1.3. Portland cement

Portland cement shall comply with BS 12:1996.

3.1.4. Ground granulated blast-furnace slag

Ground granulated blastfurnace slag (ggbs) shall comply with BS 6699:1992.

3.1.5. Pulverised fuel ash

Pulverised fuel ash (pfa) shall comply with BS 3892 Part 1:1993 and shall be from a single source.

3.1.6. Certification

The Contractor shall provide the Engineer on a regular basis with certification for constituent materials of the slurry. Such certificates shall comply with the relevant standards listed above.

3.1.7. Admixtures

If the slurry design includes admixtures, the admixtures shall have been successfully used on previous cut-off wall contracts, or their effects shall be demonstrated in trial mixes prior to the commencement of the slurry trench works.

3.2. Geomembrane

3.2.1. Geomembrane material

The geomembrane shall be high density polyethylene (HDPE) or such other material as permitted by Clause 1.4, produced from new polyethylene resins containing no fillers, plasticisers or additives other than carbon black. Antioxidants and inhibitors may also be used in small amounts not exceeding 500 parts per million. The material properties shall conform to the National Sanitation Foundation (NSF) Standard 54-1993 or later, and shall meet the additional requirements of Table 1 of this Specification. The manufactured sheet shall be smooth with a pore-free surface, and be unlaminated and free from blemishes, abrasions or other surface defects.

The geomembrane shall not be adversely affected, visually or physically, by aqueous salts and acidic and alkaline solutions within the pH range 4 to 13 (inclusive), oils, greases, common organic

Table 1. HDPE geomembrane specification requirements over and above NSF-54 requirements

Properties	Specification	Test method
Thickness	2.0±10% mm	ASTM D1593
Environmental stress crack resistance	Transition time >100 hours	ASTM D5397*
Oxidation induction time	>100 mins	ASTM D3895, 200°C, pure oxygen at 1 atmosphere
Chemical compatibility	see Particular Specification	EPA 9090

* Complete test required.

solvents and tars. In addition, the material shall be resistant to attack by materials identified in the Particular Specification according to the test procedures set out therein. The geomembrane shall be resistant to temperatures varying between 0 and 70°C. Documentation demonstrating independent verification of these requirements shall be submitted by the Contractor for approval prior to the ordering of the geomembrane.

3.2.2. Manufacturer's quality control

Each roll of geomembrane delivered to site shall be 'Quality Assured by the manufacturer and certified in accordance with systematic production quality control testing programmes. This certificate shall list the following minimum information:

(a) manufacturer
(b) product name
(c) place of manufacture
(d) date and time of manufacture
(e) production line and shift
(f) roll identification number
(g) thickness — ASTM D1593
(h) carbon black content — ASTM D1603
(i) carbon black dispersion — ASTM D3015
(j) density — ASTM D1505A
(k) tensile properties: stress at yield, stress at break, elongation at yield, elongation at break — ASTM D638

The above testing procedures are the suggested methods; however, other test methods may be used, provided that the manufacturer can demonstrate that the methods are comparable to those listed above.

3.2.3. Delivery, handling and storage

The geomembrane shall be delivered to the panel fabricating yard in rolls, handled and stored strictly in accordance with the manufacturer's requirements and recommendations. All necessary care shall be taken to ensure no damage occurs to the rolls during the panel fabrication process.

Once the fabricated geomembrane panels arrive on site, the Contractor shall provide appropriate machinery for lifting and transporting. Machine handling should be undertaken using procedures recommended by the manufacturer. Direct handling of rolls with fork-lift trucks or machine buckets, blades, etc. shall not be permitted.

3.2.4. Contractor's experience/ manufacturer's experience

The geomembrane installation Contractor shall have relevant experience of installing geomembrane of the type used in the slurry trench works. The Contractor shall submit a summary of the experience of the proposed installation Contractor with the proposed installation method to the Engineer at Tender stage.

The Contractor shall also provide prior to commencement of the slurry trench works, a summary of the experience of the Lead Technicians who will be responsible for the slurry trench works at both the fabrication yard and the job site.

The Contractor shall not change geomembrane manufacturer or installation personnel without the written approval of the Engineer.

3.2.5. Welded joints

All welded joints shall conform to the methods detailed in the US Environmental Protection Agency Technical Guidance Document, *The fabrication of polyethylene FML field seams* (No. EPA/530/SW-09/069, September 1989), or later edition. All welded joints shall be monitored throughout their length by the Engineer or appointed representative.

The Contractor shall perform trial welds with each welding machine and operator at least at the start of each shift and also following any period of machine shutdown or change of operator. Trial welds shall be at least 2 m long in the case of extrusion welds and at least 3.5 m long in the case of fusion welds. On completion of the trial weld, the Contractor shall cut a total of four 25 mm wide field tabs normal to the weld and subject them to qualitative destructive testing. Three specimens shall be tested in peel mode and one in shear mode.

The weld will be deemed to have passed the trial if the failure occurs solely in the parent material and does not enter the weld. If a trial weld fails destructive testing then the welding machine and the operator shall not be allowed to carry out welded joints until the deficiencies are corrected and both machine and operator have achieved passing trial welds. Trial welding and destructive testing shall be observed by the Engineer or his appointed representative.

The Contractor shall perform welded joints only after satisfying trial weld conditions as specified above. All pre-treatment measures (e.g. grinding and cleaning) specified in EPA/530/SW-89/069 shall be carried out, and extrudate and/or wedge temperatures shall be maintained within a range approved by the Engineer or his appointed representative.

Prior to commencement of the slurry trench works, the Contractor shall provide the Engineer with the following information:

(a) welding technique or techniques and their proposed applications
(b) welding machinery
(c) overlap widths and sheet preparation
(d) optimum temperature range for extrudate and/or hot wedge
(e) optimum welding speed of automatic weld equipment
(f) non-destructive test methods and scheme for testing.

3.2.6. Interlocking joints

The Contractor shall provide the Engineer with details of the proposed method of interlocking panels to show that this will not compromise the strength of the barrier. In particular, destructive test results are required to demonstrate that the strength of the interlocking system is not less than the tensile strength of the geomembrane parent sheet. Permeability requirements and method of testing the interlock system, including any material used in it, will be as described in the Particular Specification.

All interlock systems, as distinct from the sealing material, shall be manufactured from HDPE or such other material permitted by Clause 1.4. The Contractor shall demonstrate compliance by supplying results of density testing to the ASTM D1505A standard, and resin certificates of the material used.

The Contractor shall not change the interlocking system without the written approval of the Engineer.

4. Slurry production and quality control

4.1. Slurry preparation plant

Mixing of slurry shall be carried out in a two-stage process (unless otherwise approved by the Engineer). Bentonite powder and water shall be thoroughly mixed using high shear or colloidal mixing techniques and the bentonite mud allowed a minimum of 8 hours to hydrate (unless the Contractor can demonstrate that a shorter duration is acceptable) prior to its use in the slurry mix.

4.2. Batching

4.2.1. Accuracy

All parts of the batching plant shall be capable of weighing and/or measuring the quantity of each individual component material to an accuracy of ±3% of the target quantity of that material to be used in the design mix.

4.2.2. Calibration

The Contractor shall ensure that the batching plant measurement systems are fully calibrated prior to commencement of production for the permanent works, and thereafter at least weekly. The Engineer shall be invited to witness the plant calibrations and copies of the record of calibrations shall be forwarded to him. If any part of the plant does not meet the batching accuracy criteria work shall not proceed until it has been adjusted and recalibrated to meet the specified accuracy.

4.3. Quality control

If so required by the Particular Specification the Contractor shall inform the Engineer of the minimum viscosity and specific gravity of the bentonite suspension to be used.

4.3.1. Bentonite suspension

4.3.1.1. Sampling

Bentonite suspension shall be sampled from the mixing plant at least three times during production days. The samples shall be tested for rheological properties and specific gravity in accordance with Clauses 4.3.1.2 and 4.3.1.3. The age of the fluid at the time of testing and location where sampled shall be recorded.

4.3.1.2. Viscosity

The viscosity shall be measured by either the Marsh funnel or the direct reading viscometer.

   (i) Marsh funnel. The viscosity of the bentonite suspension shall be measured using the Marsh funnel in accordance with Appendix A.
   (ii) Direct reading viscometer. The viscosity and gel strength of the bentonite suspension shall be measured using a direct indicating viscometer in accordance with Appendix A. Readings for the instrument shall be recorded at 300 rpm and 600 rpm. The gel strength shall be recorded at 10 seconds.

4.3.1.3. Specific gravity

The specific gravity of the bentonite suspension shall be measured by the use of a fixed volume container and electronic weighing scales or by measuring the solids content of the suspension. The specific gravity shall be measured to an accuracy of ±0.002.

4.3.2. Cementitious slurry

If so required by the Particular Specification the Contractor shall inform the Engineer of the target specific gravity and bleed of the cementitious slurry to be used. The target bleed value as defined in Clause 4.3.2.3 shall not exceed 2%. In the event that the measured values differ significantly from the target values then the Contractor shall propose measures for rectification of this situation.

4.3.2.1. Sampling

Cementitious slurry shall be sampled from the mixing plant at least three times during production days. The samples shall be tested for specific gravity and bleed in accordance with Clauses 4.3.2.2 and 4.3.2.3. The age of the fluid at the time of test and location where sampled shall be recorded.

4.3.2.2. Specific gravity

The specific gravity of the cementitious slurry shall be measured by the use of a fixed volume container and electronic weighing scales such that an accuracy of ±0.002 is achieved.

4.3.2.3. Bleed

Bleed shall be measured at 24 hours after mixing using a 1000 ml measuring cylinder. The set cementitious slurry solids shall occupy a volume of at least 980 ml.

4.4. Temperature

The temperature of bentonite suspension and cementitious slurry samples taken for purposes of the quality control procedures specified in Clause 4.3 shall be recorded at the time of testing the samples.

5. Slurry wall construction

5.1. Slurry wall construction

The Contractor shall submit to the Engineer, at least two weeks prior to commencement of the slurry trench works, all relevant details of the method of construction of the cut-off wall and the proposed plant.

5.1.1. Excavation

5.1.1.1. Excavation depth

The method of excavation shall be able to obtain the specified maximum depth and accommodate changes in depth while maintaining the required penetration into the aquiclude.

5.1.1.2. Guide walls

Generally, guide walls shall be used where the trench excavation is carried out by grab techniques to provide correct alignment to the wall.

5.1.1.3. Slurry level

The working platform level shall be set such that the slurry level at all times can be kept at least 1.5 m above any groundwater level. The Contractor shall maintain the level of the fluid slurry in the trench excavation sufficiently high to minimise the risk of collapse of the trench sides. Any drop in level of the slurry due to filtration, penetration, moisture loss or any other cause shall be made up by the Contractor to ensure the as-cast level of slurry is above the required finished trimmed level.

5.1.1.4. Ground and toe levels

The Contractor shall provide the Engineer with his proposed method for dealing with:

(a) uneven ground surface to achieve the required top level
(b) variable aquiclude elevation and hence changing cut-off wall toe level to achieve the required aquiclude penetration.

5.1.2. Wall location and dimensions

5.1.2.1. Setting out

The cut-off wall shall be set out to a plan positional tolerance of ±50 mm and constructed to a plan positional tolerance of ±150 mm measured at the centre line at the top of the trench.

5.1.2.2. Width

The width of the wall shall not be less than the minimum width specified in the Particular Specification.

5.1.2.3. Verticality

The centre line of the cut-off wall shall be constructed to be generally within ±1:60 from vertical.

5.1.2.4. Toe levels, depths

The required toe level and/or penetration for the wall and geomembrane into the aquiclude is given in the Particular Specification. The depth of the wall toe below working platform level shall be recorded and verified by the Engineer's representative at chainage intervals as specified in Clause 2.1. For a wall toeing into an aquiclude, the upper surface of the aquiclude shall be identified and verified by the Engineer's representative by observation of spoil arisings and the depth to this layer as well as the toe level of the trench shall be recorded at chainage intervals as specified in Clause 2.1.

5.1.3. Day joints

If the cut-off wall is not constructed by continuous excavation, the Contractor shall adopt suitable measures to ensure that the wall is continuous across joints resulting from interruption of the excavation (e.g. day joints). Where there is no geomembrane within the wall, a minimum overdig into the previous day's work of 0.5 m throughout the depth of the wall shall be adopted to ensure intermixing at joints. The Contractor shall ensure that his excavation procedure does not cause damage to the material in completed sections of the wall.

Temporary stop-ends if used shall be of the length, thickness and quality of material adequate for the purpose of preventing water and soil from entering the completed excavations or sections of wall. Each temporary stop-end shall be rigid and adequately restrained to minimise horizontal ground movement and shall be straight and true throughout and the external surface shall be clean and smooth i.e., free from defects that could affect the wall integrity during removal of the temporary stop-end.

5.1.4. Temporary protection

Temporary protection measures to the slurry wall shall be carried out by the Contractor for that section of wall currently being excavated and thereafter until handover to the Contractor who will be carrying out the wall capping works. Any such handover will not normally be earlier than 7 days after construction of a section of the wall.

5.1.4.1. Safety

So far as is reasonably practicable, measures including notices and barriers shall be taken to prevent persons from falling into the fluid slurry or slipping on the set slurry.

5.1.4.2. Reduction of surface cracking

Measures shall be taken to reduce surface cracking of the set slurry and, if necessary, to ensure cracks do not extend below the capping level.

5.1.5. Actions on loss of slurry or collapse of trench

In the event of a sudden or sustained loss of cementitious slurry or collapse of the trench the Contractor shall take immediate action to safeguard the slurry trench works and seek instructions from the Engineer.

5.1.6. Excessive specific gravity of slurry in trench

If the slurry specific gravity exceeds the limit stated in the Particular Specification, the Contractor shall immediately inform the Engineer of the situation and his suggestion as to how to proceed, whereupon the Engineer will issue appropriate instructions to progress the slurry trench works.

5.1.7. Ambient temperature

Slurry mixing and placement may proceed so long as the Contractor is satisfied that the prevailing air, water, and mixed slurry temperatures will not be detrimental to the slurry trench works. If frozen material is present in the slurry, work shall be suspended.

5.1.8. Capping

Trimming and capping shall take place as soon as practicable to enable the capping to be placed to prevent excessive drying out of the wall.

5.1.8.1. Trimming of top of slurry

The top of the slurry wall shall be trimmed carefully from the cast level to the trimmed level so as to avoid any damage to the cut-off wall.

5.1.8.2. Capping detail

The Contractor shall provide the Engineer with his proposed method for constructing the capping detail as shown on the Drawings for approval.

5.2. Additional measures for geomembrane walls

5.2.1. Installation of geomembrane

Installation of the geomembrane shall be carried out in a manner developed by the Contractor to ensure that the geomembrane is not damaged. This may be effected by installing discrete panels with proprietary joints or by using continuous geomembrane. The system used shall ensure that full engagement and necessary alignment of any proposed joints are achieved, and that the geomembrane is positioned in the trench in accordance with the Contract requirements. Full details of the Contractor's proposed method for geomembrane installation shall be supplied to the Engineer for his approval.

A fully documented 'as installed' sheet layout plan shall be produced by the Contractor as work proceeds. The 'as installed' plan shall include as a minimum the roll positions with reference code and date placed, together with any other information requested by the Engineer.

5.2.2. Dimensions

The plan length of a geomembrane panel may be varied to suit the Contractor's equipment but any upper limits on this length due to geological or other external factors shall be as specified in the Particular Specification. Within these constraints the Contractor shall be responsible for selecting individual panel dimensions taking into account trench stability issues and the standard widths of geomembrane produced by the manufacturers. In designing alterations to the direction in plan of the slurry wall, the Engineer will take into account the need to make on site adjustments to geomembrane panel lengths to achieve directional changes.

5.2.3. Base levels, bottom details and sloping strata

Any detailing of the slurry wall requiring changes in the toe level of the geomembrane shall take into account the need to construct all individual panels with no variation in the design toe level of the geomembrane across any one individual panel.

5.2.4. Positional tolerance

Any requirement for the geomembrane to occupy a specific position within the trench will be set out in the Particular Specification together with details as to how it is to be achieved.

5.2.5. Changes in ground level

Any detailing of the slurry wall requiring changes in the top level of the geomembrane shall take into account the need to construct all individual panels with no initial variation in the design top level of the geomembrane across any one individual panel.

5.2.6. Repairs to damaged sections and joints

Any repairs carried out to damaged geomembrane sections and joints shall be carried out in accordance with the geomembrane manufacturer's recommendations and shall not compromise the quality and integrity of the finished product. Records of repairs shall be kept by the Contractor and submitted to the Engineer by the time of completion of the slurry trench works.

5.2.7. Temporary stop ends

In addition to satisfying the requirements of Clause 5.1.3, temporary stop ends where used shall adequately protect the installed geomembrane joint from becoming blocked with set slurry to allow insertion of subsequent geomembrane panels.

5.2.8. Temporary fixing and protection of geomembrane

Temporary fixing and protection of the geomembrane shall be carried out by the slurry wall Contractor for that section of wall currently being installed and thereafter for a period of seven days. Seven days after installation of a part of the slurry wall the responsibility for providing protection measures will rest with the Contractor carrying out the wall capping works.

5.2.9. Contractor's supervision

The Contractor shall ensure that constant supervision of the geomembrane installation process and any welding operation is maintained throughout the slurry trench works by the approved Site Supervisor/Lead Technician.

5.2.10. Good practice

The Contractor shall ensure that current good practice is adhered to throughout the slurry trench works and in particular he shall ensure the following:

(a) No smoking shall be allowed in the vicinity of the geomembrane.
(b) No articles are to be dragged across the geomembrane.
(c) Only flat soft-soled footwear is allowed on the geomembrane.
(d) An adequate supply of suitable cleaning materials is to be available at all times for seam preparation etc.
(e) Where the geomembrane is laid out on site prior to installation, the ground should be cleared of any sharp protrusions, stones etc.

6. Compliance testing for material properties

6.1. Sampling for testing for set properties

Samples for later testing of set properties of the cementitious slurry shall be made by pouring the representative slurry sample from the trench into sample tubes prior to the slurry setting. The number of sample tubes to be filled to make one representative sample set shall be as stated in the Particular Specification, but shall not be less than three.

6.1.1. Sample tubes

Sample tubes shall be 100 mm nominal diameter PVC tubing of minimum length 300 mm. The ends of the sample tubes shall be square and free from burrs and the inside of the tubes shall be clean. Filled sample tubes shall be labelled with the contract name, location, date and depth of sample, and given a unique reference number.

6.1.2. Frequency

Sample location points shall be at intervals of chainage along the wall such that there is one sample location point per 200 m^2 of projected wall area in elevation, and not less than one sample location per day's production. Samples must be taken during the same day as the slurry is produced, except when excavation to full depth takes more than one day.

6.1.3. Location

Representative sample sets of slurry shall be taken from one metre below the top and one metre above the bottom of the trench at each sample location point. The means of taking the samples shall ensure that the sample is taken from the zone intended.

6.1.4. Care of samples

The fluid slurry shall be carefully poured into the sample tubes in such a way that air is not entrapped in the slurry. The sample tubes shall be closed without delay by close-fitting PVC push-on end caps and sealed to avoid moisture loss. Samples shall not be transported until at least 14 days old. During storage and later transportation care must be taken to ensure that the samples are not damaged by impact or vibration. Samples shall not be demoulded until required for test.

6.1.5. Storage and transportation

The samples shall be stored in an environment where the temperature is $20°C \pm 2°C$. Samples must be stored and transported upright at all times.

6.2. Laboratory testing

Testing the set properties of the slurry shall only be carried out by an approved laboratory with suitable equipment and experience in dealing with cementitious slurries.

6.2.1. Approved laboratory

The Contractor shall submit details of the proposed testing laboratory to the Engineer for his approval at least two weeks prior to the first test commencing.

6.2.2. Scheduling of tests

Within 24 hours of the sampling, the Contractor shall submit to the Engineer a record of all samples taken giving the chainage/location, date, depth and reference number of the samples. The Engineer will provide the Contractor with a testing schedule at least two weeks before the tests are required to be carried out. Unless otherwise stated in the Particular Specification, the minimum testing requirement shall be two strength and two permeability tests for each 1000 m^2 of wall.

6.2.3. Preparation and care of samples

The samples shall be carefully extruded vertically from the sample tubes and trimmed in a progressive manner to avoid shearing the material. Any hollows in the top or bottom of the sample shall be removed by careful trimming off the surrounding high zones. Material shall not be filled back into hollows. There shall be no sub-sampling of smaller diameter samples from the original sample.

6.2.4. Tests

6.2.4.1. Triaxial permeability tests

Specified triaxial permeability tests shall be carried out and reported in accordance with the procedures stated in Appendix B.

6.2.4.2. Unconfined compression strength tests

Specified unconfined compression strength tests shall be carried out and reported in accordance with the procedures stated in Appendix B.

6.2.4.3. Reporting of results

The Contractor shall provide the Engineer with a full copy of each laboratory test result within one week of completion of the test.

6.3. In-situ testing

Where required by the Engineer, in-situ testing shall be carried out in accordance with the Particular Specification.

6.4. Geomembrane
6.4.1. Geomembrane conformance testing

As soon as practicable after the delivery of the geomembrane rolls to the fabrication yard, the Contractor shall cut and label samples 1 m wide across the entire width of the selected rolls for conformance testing and/or retention as directed by the Engineer or his appointed representative.

The Contractor will be required to submit one, or more if specified in the Particular Specification, of the samples cut from the geomembrane to an independent geosynthetic test laboratory approved by the Engineer for conformance testing. The parameters listed in Table 2 shall be tested. All test results shall be reported to the Engineer immediately on receipt and within seven days (excluding the test duration period) of the samples being taken, unless otherwise approved by the Engineer.

Failure to meet the minimum requirements set out in Table 2 in any respect may be cause for rejection of that material. Any repairs, replacement or other works occasioned by the failure of the geomembrane to meet these requirements shall be carried by the Contractor. The Contractor will be offered the opportunity to prove that other material from the rejected roll does meet the minimum specification requirements, with any further testing being at the Contractor's expense.

Table 2. HDPE geomembrane conformance testing requirements

Properties	Specification	Test method
Thickness	2.0±10% mm	ASTM D1593
Density	minimum 940 kg/m^3	ASTM D1505A
Carbon black content	>2% by mass	ASTM D1603
Carbon black dispersion	A1, A2 or B1	ASTM D3015
Tensile properties:		ASTM D638, Type IV, 50 mm/min
stress at yield	>15 N/mm^2	
elongation at yield	>12%	
stress at break	>26 N/mm^2	
Tear resistance	>200 N	ASTM D1004, Die C
Puncture resistance	>200 N/mm	FTMS 101C, Method 2065
Environmental stress crack resistance	>200 hours	ASTM D5397, single point test, 30% of the room temperature yield stress

6.4.2 Geomembrane non-destructive testing

All joints welded at the fabrication yard or on site shall undergo non-destructive testing in accordance with the Contractor's agreed proposals, over their entire length. Prior to commencement of the slurry trench works, the Contractor shall issue a method statement detailing his proposed non-destructive testing techniques and their proposed application. Suitable test methods include pneumatic, ultrasonic, electrical spark or vacuum box methods.

Welded joints that fail the test method must be noted and clearly recorded. Each failure shall be immediately repaired by a method agreed between the Contractor and Engineer, and the repair shall then be subjected to further non-destructive testing until the repair passes. No geomembrane shall be placed into the trench until all welded joints have passed non-destructive tests, approved by the Engineer.

6.4.3. Geomembrane qualitative destructive testing

If required, geomembrane qualitative destructive testing shall be carried out in accordance with the procedures set out in the Particular Specification.

If failure occurs through poor welding procedures the Contractor shall submit to the Engineer his proposals for remedying the welding procedure.

Should failure occur within the interlock then the Contractor shall propose another interlocking system for the slurry trench works.

6.4.4. Geomembrane quantitative destructive testing

Wide width samples of the completed welded interlock/geomembrane sheet joint may be required in the Particular Specification, and subjected to laboratory quantitative destructive testing. The required strength of the interlock bond shall be stated in the Particular Specification.

6.4.5. Geomembrane interlock performance

Where required by the Engineer, the geomembrane interlock performance in terms of flow per unit length shall be specified and demonstrated using the test method stated in the Particular Specification.

7. Records

7.1. Slurry wall records

Records of the construction of the slurry trench works shall be made by the Contractor during the installation of the cut-off wall and these shall be submitted to the Engineer for his signature. The records shall as a minimum include:

(a) Daily testing and sampling report
(b) Daily slurry trench report
(c) Weekly slurry sample register
(d) Date of shift
(e) Names of key personnel and their duties
(f) Names of any key geomembrane installation team personnel and their duties
(g) Weather conditions including maximum and minimum daily temperatures
(h) Daily progress, trench excavation, any geomembrane installation including any welding, toe-in, trench lengths to date, fault record, testing results, etc.

Typical examples of the information required in these forms are given in Appendix C.

7.2. Additional records for geomembrane walls

7.2.1. 'As-installed' plan

The Contractor shall prepare an 'as-installed' plan for the geomembrane panels as described in Clause 5.2.1 above. The final plan shall be submitted to the Engineer within seven working days of the completion of the slurry trench works.

7.2.2. Geomembrane fault plan

A record in the form of a fault plan shall be maintained by the Contractor and submitted to the Engineer on completion of the cut-off wall. The fault plan shall include as a minimum:

(a) material production faults
(b) mechanical damage
(c) weld discontinuities
(d) jointing problems
(e) details of remedial works carried out.

8. Post-construction monitoring

The Particular Specification will state any post-construction monitoring requirements.

Appendix A. Tests for bentonite and slurry fluid properties

A.1. Marsh funnel viscosity test

Apparatus and testing method shall be generally to American Petroleum Institute Recommended Practice RP 13B-1 as follows.

A.1.1. Equipment

The equipment consists of the following:

(a) Marsh funnel
A Marsh funnel is calibrated to out-flow one US quart (946 ml) of fresh water at a temperature of 70±5°F (21±3°C) in 26±0.5 seconds. A graduated cup is used as a receiver.[1]

Specification
Funnel cone
    Length   12.0 in (305 mm)
    Diameter (top)   6.0 in (152 mm)
    Capacity to bottom of screen   1500 ml
Orifice
    Length   2.0 in (50.8 mm)
    Inside diameter   3/16 in (4.7 mm)
Screen   12 mesh US
    Has 1/16 in (1.6 mm) openings and is fixed at a level 3/4 in (19.0 mm) below top of funnel.

(b) Graduated cup: one US quart (946 ml)[1]
(c) Stopwatch
(d) Thermometer: 32–220°F (0–105°C)

A.1.2. Procedure

The procedure is as follows.

(a) Cover the funnel orifice with a finger and pour bentonite suspension through the screen into the clean, upright funnel. Fill until fluid reaches the bottom of the screen.
(b) Remove finger and start stopwatch. Measure the time for mud to fill to one-quart mark of the cup.
(c) Measure temperature of fluid in degrees C.
(d) Report the time to nearest second as Marsh funnel viscosity. Report the temperature of fluid to nearest degree C.

A.2. Direct reading viscometer

Apparatus and testing method shall be generally to American Petroleum Institute Recommended Practice RP 13B-1 as follows:

A.2.1. Equipment

The equipment consists of the following:

(a) A direct-indicating viscometer powered by an electric motor or hand crank e.g. Fann Viscometer. Bentonite suspension is contained in the annular space between two concentric cylinders. The outer cylinder or rotor sleeve is driven at a

---

[1] One litre may be used instead of 946 ml, in which case the appropriate calibration time should be used and the result should clearly reference the one litre volume.

constant rotational velocity (rpm). The rotation of the rotor sleeve in the fluid produces a torque on the inner cylinder or bob. A torsion spring restrains the movement of the bob, and a dial attached to the bob indicates displacement of the bob.

Instrument constants have been adjusted so that apparent viscosity, plastic viscosity and yield point are obtained by using readings from rotor sleeve speeds of 300 rpm and 600 rpm.

Specifications: Direct-indicating viscometer
Rotor sleeve
Inside diameter	1.450 in (36.83 mm)
Total length	3.425 in (87.00 mm)
Scribed line	2.3 in (58.40 mm) above the bottom of sleeve

Two rows of 1/8 in (3.18 mm) holes spaced 120 deg (2.09 radians) apart around rotor just below scribed line.

Bob
Diameter	1.358 in (34.49 mm)
Cylinder length	1.496 in (38.00 mm)

Bob is closed with a flat base and tapered top.

Torsion spring constant
386 dyne-cm/degree deflection
Rotor speeds
High speed	600 rpm
Low speed	300 rpm

(b) Stopwatch.
(c) Suitable container, e.g. the cup provided with viscometer.
(d) Thermometer: 32–220°F (0–105°C).

A.2.2. Procedure

The procedure is as follows.

(a) Place a sample in container and immerse the rotor sleeve exactly to the scribed line. Measurements in the field should be made with minimum delay (within five minutes, if possible from sampling) and at a temperature as near as practical to that of the mud at the place of sampling (not to differ more than 10°F (6°C)). The place of sampling should be stated on the report.
(b) Record the temperature of the sample.
(c) With the sleeve rotating at 600 rpm, wait for dial to reach a steady value (the time required is dependent on the mud characteristics). Record the dial reading for 600 rpm.
(d) Shift to 300 rpm and wait for dial reading to reach steady value. Record the dial reading for 300 rpm.

A.2.3. Calculation

Plastic viscosity (in cP) = 600 rpm reading − 300 rpm reading

$$\text{Apparent viscosity (in cP)} = \frac{600 \text{ rpm reading}}{2}$$

Yield point (in lb/100 ft^2) = 300 rpm reading − plastic viscosity

A.3. Specific gravity test
A.3.1. Equipment

The equipment consists of the following.

(a) A set of calibrated weighing scales of sufficient accuracy to permit measurement of weight to within ±0.1 g.
(b) A fixed volume container with a flat upper edge generally no smaller than 1.0 litre capacity with a closely fitting top plate.

A.3.2. Procedure

The procedure is as follows.

(a) The empty fixed volume container and the top plate is carefully dried and weighed and the weight is recorded, W_a.

(b) The fixed volume container is filled to the top with water. The top plate is carefully slid across the top of the container to exclude all air bubbles. Any excess water is removed to avoid errors. The container, top plate and water are carefully weighed and the weight recorded, W_b.

(c) The fixed volume container is emptied, dried and filled to the top with fluid to be tested. The top plate is carefully slid across the top of the container to exclude all air bubbles. Any excess fluid is removed to avoid errors. The container, top plate and fluid are carefully weighed and the weight recorded, W_c.

(d) The specific gravity of the fluid is calculated as follows
specific gravity $= (W_c - W_a)/(W_b - W_a)$.

(e) The accuracy of the weighing scales should be checked at least weekly.

A.4. Bleed test
A.4.1. Equipment

The equipment consists of a plastic or similar material vertical measuring cylinder graduated in 20 ml increments up to 1000 ml, and a membrane or similar to seal the measuring cylinder top to avoid evaporation.

A.4.2. Procedure

The procedure is as follows.

(a) Carefully fill the measuring cylinder to the 1000 ml mark with the cementitious slurry, ensuring that no air is entrapped. Seal the top of the measuring cylinder. Leave to stand undisturbed for 24 hours.

(b) When the slurry has stood for 24 hours record the volume occupied by the set solids.

Appendix B.   Tests for set slurry properties

B.1. Determination of permeability in triaxial cell

The test is in accordance with BS 1377: 1990 Part 6, Clause 6, at the prescribed age.

The test conditions are:

(a) specimen 100 mm diameter by 100 mm long minimum
(b) direction of flow along length of specimen, vertically upwards
(c) saturation according to BS 1377:1990 Part 6, Clause 5.4.3 (where necessary)
(d) effective confining stress of 100 kPa
(e) no requirement for void ratio to be calculated.

In addition the following conditions shall apply:

(f) the specimen must not be sealed in the cell by wax.
(g) the hydraulic gradient shall be between 10 and 20 as scheduled by the Engineer and the specimen allowed to equilibrate for at least 12 hours before flow measurements are taken. Thereafter, flow shall be measured over a period of 48 hours.
(h) there shall be no equipment used which is sensitive to alkaline solutions.

Results to be reported as coefficient of permeability (m/s) corrected to 20°C.

B.2. Determination of the unconfined compressive strength

The test is in accordance with BS 1377:1990 Part 7, Clause 7. The test conditions are:

(a) specimen 100 mm diameter by 200 mm long.
(b) strain rate to be 0.1% per minute.

Results to be reported as (kPa).

Appendix C. Typical example report forms

TYPICAL EXAMPLE DAILY TESTING AND SAMPLING REPORT

CONTRACT: _____ CONTRACT NUMBER: _____ DATE: _____ SHEET NO: __ OF __

Bentonite: **kg/batch**
Cement:
Pfa/ggbs:
Water:
Additive: **litres**

	BENTONITE SUSPENSION AND SLURRY PRODUCTION CONTROL TESTS				
	BENTONITE SUSPENSION				FRESH CEMENTITIOUS SLURRY
	Fresh		Hydrated		from Mixing Plant
Time of mixing					
Time of sampling					
Time of testing					
Viscosity (Marsh Funnel) (s)					N/A
Bleed set solids/1000ml	N/A	N/A	N/A	N/A	
Viscometer readings 300/600	/	/	/	/	N/A
Specific gravity					
Slurry Temperature °C					
Notes					

SLURRY COMPLIANCE TESTING/SAMPLES	
	CEMENTITIOUS SLURRY SAMPLED FROM TRENCH
Sample set No	
Chainage (m)	
Time (am/pm)	
Depth Sampled (m)	
Notes eg degree of contamination etc	

NOTES, INSTRUCTIONS RECEIVED, ADDITIONAL SAMPLES REQUESTED ETC

_____
_____
_____
_____
_____
_____
_____
_____

SIGNED: _____   Distribution: CLIENT
                                                    CONTRACTS MANAGER
                                                    QS DEPT
                                                    SITE

Note N/A: although these tests are not required by this specification, some contractors may offer these tests.

TYPICAL EXAMPLE DAILY SLURRY TRENCH REPORT

SHEET NO ___ OF ___

CLIENT: _____

CONTRACT NAME: _____ CONTRACT NUMBER: _____

DATE EXCAVATED: _____ BANKSMAN: _____

EXCAVATOR: _____ GRAB/BUCKET WIDTH (a): _____ m

STRATA RECORD		EXCAVATION RECORD AT CHAINAGE INTERVALS OF ___ m (see clause 2.1)			
CHAINAGE		CHAINAGE	DEPTH TO AQUICLUDE (m)	WALL TOE DEPTH (m)	ENGINEER'S REPRESENTATIVE SIGNATURE
DEPTH	DESCRIPTION				
		AVERAGE DEPTH (c)			

START CHAINAGE: _____ TIME: _____

FINISH CHAINAGE: _____ TIME: _____

EXCAVATED LENGTH (b): _____ (m)

THEO. VOL. $d = (a \times b \times c)$: _____ m³   SLURRY VOL BATCHED: _____ m³ (e)

DAILY OVERBREAK $(e-d)/(d)$: _____ %

NOTES (Obstructions, Instructions, Breakdowns, Weather Conditions etc) _____

_____

_____

_____

_____

Signed: _____ CLIENT   Distribution: CLIENT
                                                          CONTRACTS MANAGER
Signed: _____   QS DEPT
   Note: all dimensions in m unless noted otherwise   SITE

TYPICAL EXAMPLE WEEKLY SLURRY SAMPLE REGISTER

CONTRACT NAME: _____ CONTRACT NUMBER: _____ WEEK ENDING: _____ SHEET NO: ___ OF ___

SAMPLE NO	DATE MADE	CHAINAGE (m)	SAMPLE DEPTH (m)	SAMPLE SEALED DATE	ENGINEER'S TEST SELECTION					SAMPLE TO LAB	SAMPLE TO STORE	SAMPLE TO OTHER	NOTES
					PERMEABILITY	AGE	UCS	AGE					

COMPILED: _____

DISTRIBUTION:

FOR TEST SELECTION:
RE
SITE FILE

AFTER TEST SELECTION:
CONTRACT MANAGER
TEST LAB (with sample)
QS DEPT
SITE FILE

TEST SELECTION INSTRUCTED: _____   DATE: _____
                            RESIDENT ENGINEER

Specification for the construction of slurry trench cut-off walls

Notes for guidance for the construction of slurry trench cut-off walls
as barriers to pollution migration

The Notes for Guidance have been written to develop and expand the Specification. They are for guidance only and do not form part of the Specification. A bibliography of relevant references, standards and test methods is included.

Introduction—General issues

The following are some general points relating to cut-off walls which are not specifically addressed in the Specification:

Hydraulic conductivity/ permeability/ permittivity

The hydraulic conductivity, k (m/s) of a material is defined by Darcy's law as:

$$k = q/iA$$

where q is the flow rate through the material (m^3/s), i is the hydraulic gradient across it (m/m) and A is the area of flow (m^2).

The hydraulic conductivity defined according to Darcy's law will have units of length/time and in the UK it is common practice to use m/s. Darcy's law does not include any parameters to describe the properties of the permeant and in particular its viscosity and density. It is assumed to be pure water (or that the fluid used in any testing, e.g. a saline water, is the same as that which occurs in-situ). Thus the viscosity and density may be taken as constants though it may be necessary to correct for differences in temperature, for example, between laboratory and field situations. These properties of the permeant can be included by considering the specific permeability K (m^2) of the material. This is defined as:

$$K = k\eta/\rho g$$

where η is the viscosity of the permeant (mPas, millipascal second), ρ is the density of permeant (Mg/m^3) and g is the acceleration of gravity (m/s^2).

In theory, K is a property of a material, expressed as an area, and is independent of the permeant. However, for many cut-off materials such as clays and cement-bentonites there will be some interaction between the permeant and the material and K will show considerable variation, for example between oils and water, fresh and saline waters and liquids and gases.

The above definitions are given for completeness. In the civil engineering literature hydraulic conductivity and permeability are often taken as synonymous and the term permeability is the more generally used. This practice has been followed in the Specification and Notes for Guidance and the term permeability is used throughout rather than hydraulic conductivity.

It is also useful to define the term permittivity. The effectiveness of a barrier in controlling fluid permeation will depend on its thickness as well as its permeability to the fluid(s) which it is required to retain. Thicker walls of lower permeability will give better control of flow than thinner more permeable walls. It is the permeability divided by the thickness that is the true measure of the effectiveness of a barrier to permeation, i.e. its hydraulic resistance. This parameter is known as the permittivity of the barrier. Thus, a barrier of permeability 1×10^{-9} m/s and thickness 0.8 m would have a permittivity of 1.25×10^{-9} s^{-1}, and a 2 mm thick geomembrane would have to have a permeability of 2.5×10^{-12} m/s to achieve the same permittivity.

The concept of permittivity can be useful when assessing multiple barrier systems, for example, for cement-bentonite-geomembrane composite barriers and it can be useful to specify the barrier in terms of its overall permittivity (or the equivalent permeability of a wall of the same overall thickness). Thus, a cement-bentonite wall with a geomembrane might be required to show an equivalent permeability over its full thickness of 1×10^{-9} m/s. However, in order to evaluate this overall permeability, it will be necessary to consider the composite action of the cement-bentonite and not just the permeabilities and thicknesses of the individual elements. It will be necessary to consider the performance of the joints between geomembrane panels/sheets and any special environmental conditions, e.g. drought which, by drying a cement-bentonite material, could increase the permeability of the cement-bentonite to gas. Also it should be remembered that different components of a barrier may have quite different levels of effectiveness against different contaminants. For example, an HDPE (high density polyethylene) geomembrane may be effectively impermeable to inorganic salts such as sodium chloride but may present only a limited barrier to the diffusion of organics such as chlorinated solvents.

Gas permeability

When saturated with water, a cement-bentonite material may be effectively impermeable to gas. Any gas released at the downstream face of a water-saturated wall would have moved either in solution in the pore water or have been generated by degradation of organics which have passed through the wall in solution. In order to establish free-phase gas flow through the saturated material a substantial pressure is needed to force the pore water from the capillaries of the material and establish gas flow paths. However, if there is drying and sufficient water is lost to establish a continuous gas-filled void system within the material, a cement-bentonite material can become significantly permeable to gas. A geomembrane therefore should be included in all walls intended for the control of gas migration unless it can be demonstrated that the cement-bentonite will be sufficiently impermeable under all operating and seasonal conditions.

The role of the wall

The fundamental role of the slurry trench walls considered in this document is to act as mechanical barriers to pollution migration by control of the movement of water (clean or contaminated) and/or gas or vapours in the ground. Cement-bentonite materials may also retard the migration of some species as a result of their high pH due to the cement (some heavy metals are precipitated at high pH). Migration control by interaction of barrier and permeant may be referred to as active or reactive containment. This aspect of slurry trench cut-off walls is not considered in this Specification and the reader should refer to specialist publications.

For this Specification permeability is the fundamental parameter and many problems relating to the Specification could be resolved if there were a satisfactory test procedure for the measurement of in-situ permeability of wall sections. To date attempts based on core sampling or the insertion of probes have had only moderate success. Test boxes have been constructed contiguous with walls and have proven permeabilities to the less than 1×10^{-8} m/s level and to the less than 1×10^{-9} m/s level though the latter has been markedly more difficult. Alternative procedures currently being evaluated include:

- the insertion of pressure monitoring probes into the wall,
- precise analysis of pressure responses in boreholes adjacent to a

wall following the injection of water at a controlled flow rate for a defined period into a borehole on the opposite side,
- insertion of a piezocone.

The subject of in-situ permeability measurement is still developing for cut-off materials and constructed walls. Therefore, procedures set out in the Specification should not be regarded as invariant.

The role of the slurry in geomembrane walls

In geotechnical containment systems such as landfill liners it is common practice to use multiple barrier elements as the performance of the combined system may be markedly better than that of any of the individual components, which in the operating environment can have imperfections. For cement-bentonite-geomembrane systems the benefits may include improved permeability to gases under drought conditions and higher strain at failure if the wall is subject to movement after the cement-bentonite material has hardened.

The designer should decide what is required of the slurry when the wall includes a geomembrane, and examine the mechanisms by which the system can be compromised and procedures to eliminate or overcome these mechanisms.

Hydrogeological survey

In addition to appropriate site investigation it will be necessary to carry out a hydrogeological survey of the area in which the cut-off wall is to be sited. This survey should include consideration of the permeability of the various geological units in the area of the cut-off wall or which it may penetrate, groundwater levels including their seasonal variations, the possibility of artesian conditions and the effect of rainfall, etc. In considering the performance of a cut-off wall it will be necessary to consider not only the flow through the wall itself but also infiltration/evaporation at the surface of the site, inflow/outflow through the base of the contained/affected area, discharges from surface water drains and the possibility of leaking services, etc. Changes in the use of the site following development also may have to be considered if information is available.

It should be noted that the contained region within a cut-off wall can become the equivalent of a pond. If ponding is a possibility then drainage or other control will be needed to remove water from the contained area. The water removed from a contained area may have to be treated before disposal. This will increase the running cost of the site, but if the capping system is effective the quantities can be modest. Leachate removed by drainage will contribute to a slow clean-up of the mobile and thus potentially the most hazardous contamination, which otherwise could remain as a permanent and unchanging risk within a contained site. If a contained but uncapped site does not slowly pond with infiltrating rainwater it may be appropriate to consider whether the base and/or the cut-off is appropriately impermeable.

Future developments

The Specification is for cement-bentonite and cement-bentonite-geomembrane systems, as these are currently widely used in the UK. However, these are not the only wall types available. Other systems include: plastic concretes, silicate/silane/aggregate mixes and mixed systems containing both sodium and calcium bentonite. New materials and testing methods will be developed, and as these become available new forms of specifications and/or modifications to this Specification will be necessary.

Notes relating to clauses in the Specification

These Notes for Guidance explain or amplify the thinking behind the related sections of the Specification. Some topics occur regularly throughout the Specification, and where appropriate extended notes have been provided in one section and cross-referenced as necessary.

1. General requirements for slurry trench cut-off walls

1.1. Standards and definitions

The Specification requires that all materials and workmanship should be in accordance with appropriate British Standards. However, it is recognised that cement-bentonites are a distinct class of materials and that specifications relating to cement or geotechnical testing may not be ideal. The development of more appropriate standards would be welcomed.

1.2. General contract requirements

The particular nature of cement-bentonite materials impacts not only on the specification of materials but also on the contract itself. For example, compliance testing for cement-bentonite properties such as permeability generally will be carried out at a sample age of 90 days or greater. By this time the slurry trench construction works on a site may be complete and the contractor may have demobilised. The age at testing therefore must be considered when defining contract periods and maintenance periods.

1.3.1. Responsibility for execution and performance

Clause 1.5 of the Specification requires that a number of matters relating to the wall are described in a Particular Specification which is to be provided by the Engineer.

Two types of Particular Specification can be used for the cut-off element of a barrier wall:

- A performance Particular Specification by which the Engineer specifies the properties required of the material or end product, e.g. permeability and strength at a prescribed age and the techniques by which they are to be measured. The mix design will then be the responsibility of the cut-off wall contractor, who should be permitted an unrestricted choice of materials subject to satisfactory quality assurance and quality control measures. If this type of specification is adopted then sufficient time must be allowed within the Contract period, prior to commencement on site for the development of the mix design if properties are required which are not typical for a cement-bentonite material. There needs to be made due allowance for receipt of results from samples of 90 days' age, or other specified age. At the present time the performance specification is the norm for cut-off walls in the UK.
- A prescriptive or method specification whereby the Engineer provides complete details of the whole wall system including the mix design, mixing plant and procedures to be used by the contractor and approved suppliers.

In addition to the execution and performance of the cut-off wall it is necessary to consider its design. The contract documents will state who is responsible for the design of the slurry trench cut-off wall. It is assumed that the design will be entrusted to Chartered Engineers who have sufficient knowledge of the principles and practice of geotechnical engineering and the relevant construction materials.

1.5. Particular Specification

Clause 1.5 of the Specification sets out the matters which must be described in the Particular Specification but allows for other project specific matters to be included. Such project specific matters might include details of any in-situ testing of the wall or the inclusion of special materials in the slurry mix.

1.6. Materials

Issues relating to materials are addressed in Section 3.1 for the components of cement-bentonite mixes and Section 3.2 for geomembranes.

1.7. Site conditions
1.7.1. Topographical information

Sloping or undulating sites can present problems for slurry trench work. On sloping sites it is common practice to work uphill, and a shallow 'dam' may be inserted into the trench or other procedures used to retain the slurry in the later, higher parts of the works. If there are sharp or step changes in level it may be necessary to work downhill and to wait until the slurry in adjacent higher parts of the works has hardened sufficiently before excavation is undertaken in lower areas.

On some sites it may be necessary to excavate a wall in a series of short panels on a hit and miss basis (i.e. excavating panels 1, 3, 5, ... followed by 2, 4, 6, . . .). This type of excavation may also be necessary for deep walls or other situations where continuous excavation with a backhoe cannot be achieved.

1.7.2. Geotechnical information

As a minimum the following geotechnical information will be required:

- Information relevant to assessing the excavatability of the ground should be provided to enable the correct selection of the excavation plant for the trench. At the present time, in favourable ground conditions, walls to a depth of about 16 m (from the working platform level) may be excavated with a backhoe or similar plant. It may be noted that walls to a depth of 30 m have been excavated in the UK with a backhoe. For greater depths, or in unfavourable grounds, a cable-hung or kelly-bar mounted grab may be necessary. The choice of excavation plant may have considerable impact on the rate of excavation and also on the sequence of excavation, e.g. the adoption of semi-continuous or hit-and-miss panel excavation (see Section 5.1.1.1).
- Descriptions and thicknesses of soil strata to allow the assessment of the soil properties.
- Grading curves for the soils should be provided to give an indication of the slurry losses that may occur to the ground (coarse soils) and the potential for contamination of the slurry by fines (fine soils, see Section 5.1.1).
- Groundwater level data, including the date/seasonal conditions when obtained. The groundwater level will influence the stability of the trench, and if the groundwater level is or could be less than 1.5 m below the intended working platform level, the Engineer should consider raising the working platform level or installing a local dewatering system.

Artesian water and gas pressures

Artesian water pressures in the soil profile could severely damage the wall. It is not sufficient to ensure that the hydrostatic pressure of the fluid slurry is greater than that of the groundwater. As the slurry begins to set, the pressure it exerts will drop to no more than that of water and, indeed, during the setting process the pore pressure within the slurry can drop below the hydrostatic pressure for water. This may be particularly important in relation to the development of gas leakage paths as considered below.

The wall also may be damaged by high gas pressures, e.g. when excavating in or near landfills, or in peat. The pressure of the gas may be such that it cannot be restrained by the hydrostatic pressure of the fluid slurry or the flow resistance offered by the stiffening slurry. Once a gas flow path is established, it is unlikely to close subsequently even if the gas pressure reduces, unless gas flow ceases while the slurry is still fluid and it is re-worked. A single gas flow path can collect gas from both sides of a wall, as once a channel has been formed it may become a focus for gas flow. Gas discharges at the surface of a slurry trench should be a cause for concern, as the integrity of the wall may have been compromised unless, for example, a geomembrane has been included in the wall.

Peat

The presence of peat (or other degradable organic matter) in the soil profile through which the wall is to pass can pose a number of problems, including those listed here.

- If the peat contains gas under pressure, the problems identified above for landfill gas may occur. As a countermeasure it may be necessary to include a geomembrane in the trench or to employ some venting prior to the works, or other geotechnical measures.
- The peat could have a similar density to that of the slurry. Thus peat may tend to float from the bucket or from the excavation face or generally 'hang' in the slurry if it is not cleanly removed by the excavator. If the peat rises to the surface of the slurry or sinks to the base of the trench, it can be removed without problem. However, as the density difference between peat and slurry can be small and the slurry viscous and gelling, peat may hang in the trench. If large pieces of peat remain in the trench they could compromise its performance as a barrier. Comparable problems may occur when excavating in low density landfill or other waste materials.
- The porewater in some peats can be markedly acidic and thus damaging to cement based materials. Measurement of pH and assessment of the potential for damage may be necessary.

1.7.3. Chemical information

The chemistry of the soil, groundwater and any soil gases should be provided to the contractor so that appropriate site safety plans can be developed prior to commencement of the works.

It will also be necessary to consider the effects that such chemistry may have on the disposal of the spoil from the trench excavation and any contaminated wash water, etc.

The soil and groundwater chemistry may influence the selection of the cut-off mix and affect its durability. This is considered in Section 2.4.3.

1.8. Workmanship
1.8.2. Reinstatement

Where the cut-off wall construction is the responsibility of a specialist sub-contractor and the reinstatement that of a main contractor, there needs to be clear definition of who is responsible for temporary protection of the wall and the final capping, see Sections 2.2 and 5.1.8.

1.9. Access

Slurry trench cut-off walls may involve large quantities of bentonite, cement and ground granulated blastfurnace slag. On most jobs, these materials will be delivered by road tanker. Adequate access roads of appropriate dimensions and construction will be required to the storage silos (or the storage area if bagged materials are used). Cut-off wall construction also requires large quantities of water. This is considered in Section 3.1.1.

1.10. Working areas

The width of the working platform required for cut-off construction will depend on the plant used for the works and the procedures for spoil disposal. There should be adequate working space and access routes for the excavation plant, craneage and spoil disposal vehicles, etc.

1.11. Disposal of waste arisings

The arisings from the trench will be the excavated soil together with some adherent and fluid slurry. The high pH of the slurry can place constraints on spoil disposal and the arisings are likely to be designated controlled waste. The excavated material may also contain some contaminants from the ground and chemical analyses may be necessary to enable its disposal. Spoil disposal should meet duty-of-care guidelines and must conform to relevant regulations.

Cut-off trenches may collect a layer of bleed water, rainwater, surface water (possibly contaminated), etc. If a substantial quantity of water develops on the surface of completed sections of the wall, it should not be allowed to mix with the slurry in parts of the trench that are under excavation. If such water has to be removed to prevent intermixing, it may have to be pumped to foul sewer or disposed of by tanker. Chemical analyses may be required to enable disposal. However, a layer of water will be beneficial in preventing drying of the setting and hardening of the cement-bentonite material and thus surface water should not be removed unless there are problems such as dilution of the adjacent excavation slurry.

1.12. Setting out

See Section 5.1.2.1 of these Notes for Guidance.

2. Wall requirements

2.1. Toe depth

Cut-off walls may be required to toe into an aquiclude or to terminate in permeable ground to form a partially penetrating wall. For either type of wall the toe level has to be specified. This may be a level relative to a datum or it may be a required penetration into an aquiclude layer or other soil stratum.

Walls penetrating into an aquiclude

The depth to the aquiclude may vary and often it will have to be established by observation of the arisings. As the establishment of the correct amount of toe-in can be crucial to the proper performance of the barrier, a representative of the Engineer should be present throughout the excavation works. Once the upper level of the aquiclude has been established, the necessary toe-in can be excavated. It is for the Engineer to specify the required depth of toe-in and the procedure by which this is to be confirmed. Typical toe-in depths are 1 to 1.5 m, but depend on the function of the wall, the hydraulic head across the wall, the permeability of the adjacent strata, the nature of the aquiclude and the contaminant species to be retained, etc.

Partially penetrating walls

Cut-off walls may be used to retain floating organics or soil gases. If the concentrations of organics or gases that can dissolve in and move with the groundwater beneath a wall can be accepted, it may be possible to use a partially penetrating wall. This is a wall which extends to below the groundwater level, but not to an aquiclude. The necessary penetration of the wall into the groundwater requires careful analysis including consideration of:

- the lowest groundwater level that could occur during the operational life of the wall,
- the level of floating organics that could build up in the retained region and thus depress the level of the groundwater/organics interface,
- the depth of the cone of depression if floating organics are to be recovered by pumping from within the contained area,
- the gas pressure that could develop within the contained area and so depress the local groundwater level.

Toeing into rock

To form the toe of the wall in rock or other material that cannot be excavated with the backhoe or grab used for the main works, either chiselling or the use of a cutter-miller machine (hydromill) will be necessary. This may require either the use of a slurry with an extended set time or the use of a two-phase wall system, i.e. excavation of the wall under a bentonite slurry and subsequent backfilling of the excavation with a cut-off mix. For two-phase walls, plastic concrete is preferred to cement-bentonite slurry as with the latter the density difference between excavation slurry and backfill will not be sufficient to ensure full displacement.

If a hydromill is used spoil removal will be by slurry circulation and it will be necessary to consider the cleaning of the slurry (see Section 5.1.6).

2.2. Capping requirements

The cap is an integral part of any cut-off wall and needs careful design including the consideration of the following:

Drying of the cement-bentonite material

Drying of cement-bentonite materials is an important issue as substantial cracks can develop. These cracks will not heal on re-wetting. Cement-bentonite materials will become gas permeable if sufficient water is lost to permit the development of a continuous gas filled void system within the material.

Thus, cement-bentonite cut-off materials must be capped to prevent drying and cracking of the wall during periods of low rainfall and drought. Data on soil moisture profiles and their seasonal variation are generally sparse, but some information may be obtained from landfill cap design procedures, etc. Data on the drying behaviour of cement-bentonite materials are scarcer and in critical situations some testing would be necessary. In general, mixes of higher solids content are less sensitive to drying than those of lower solids content.

The groundwater levels on either side of the wall

The level of the groundwater on either side of the wall and the variation of these levels seasonally and over the design life of the wall must be considered. These levels are significant not only because they will influence the depth of wall that is subject to drying, but also because of the possibility for leakage at the slurry-capping interface. If the groundwater level is always going to be below the final trim level of the slurry, the purpose of the cap will be limited to the control of drying, and the provision of mechanical protection and some effective stress on the near-surface cut-off material.

Leakage at the cap/wall interface

If the groundwater level could be above the finished slurry level in the trench, the potential for a leakage path at the cap/slurry interface must be considered. It can be difficult to achieve a fully effective seal between the slurry and a clay cap because of the contrasting properties of the materials. For such situations, it may be more appropriate first to place and compact a layer of clay (or other suitable capping material) to an appropriate depth and a width of perhaps three times that of the slurry trench, and then to excavate the trench through this compacted clay. The upper section of the wall will then be contained within walls of capping material which can be extended to the necessary height to contain the groundwater. However, it should be noted that this procedure may not be suitable for use under all trench conditions and with all capping materials. The newly placed material may be unstable and collapse into the trench.

Effective stress on the near-surface cement-bentonite material

See the discussions on strain at failure in Section 2.4.2, and on expansive chemical reactions in Section 2.4.3.

Timing

The capping work should be carried out as soon as the wall has hardened sufficiently. It is important that capping proceeds with the wall excavation and is not left as part of general reinstatement at the end of wall construction. The programming of this work is important, requiring liaison between the cut-off wall contractor and the contractor responsible for the capping work. The wall should also be provided with temporary protection after excavation and prior to capping. This is considered in Section 5.1.4.2.

Trimming

Prior to capping it may be necessary to trim the wall to remove dried, cracked or other unsuitable material so as to provide a clean top surface for the capping operation.

Alignment
If required by the Engineer, markers should be set during the capping process to delineate the wall which otherwise could be obscured by the capping process or by changes at the site in ensuing years.

2.3. Fluid properties of slurry

For single-phase walls, the slurry has to support the trench during excavation and also set to form the finished wall. Important properties of the slurry are discussed below.

Specific gravity
The specific gravity of the fresh (as mixed) slurry is a function of its composition. As this includes cement, its specific gravity will be higher than that of a simple bentonite slurry. The extra specific gravity should provide additional stability to the trench. Weighting agents or fillers have been added to increase the specific gravity of a slurry, but this is not normal practice.

In use the slurry will generally increase in specific gravity as a result of suspension of material from the excavation (see Section 5.1.6).

Rheological properties
Cement-bentonite slurries are complex, time-dependent, non-Newtonian fluids. For control purposes the fluids can be characterised by an apparent viscosity, a plastic viscosity and gel strength(s) after defined rest period(s). The results obtained for the viscosity parameters will be a function of the shear rate used in any test and the geometry of the test system. The gel strength of simple bentonite slurries is reversible and develops quite markedly on standing. It is standard practice to measure the gel strength of these slurries at rest periods of 10 seconds and 10 minutes after high shearing. For cement-bentonite slurries, although they are setting materials, the 10-second and 10-minute gel strengths may be little different. Indeed, it is not unusual for the 10-minute value to be less than the 10-second value perhaps as a result of slight segregation of water and solids in the test instrument. Both values can be difficult to determine and tests will show poor repeatability as the slurry can be rather viscous and the gel strength is not sharply defined. Gel strength is therefore not recommended as control parameter for cement-bentonite slurries and the test procedure is not included in the procedure for the direct reading viscometer detailed in Appendix A, Clause 1.2 of the Specification.

The rheological properties of a slurry are sensitive to the age of the slurry at the time of test and to its shear history and particularly its immediate past shear history (standing, sheared in a mixer, passed through a pump etc.). For example, conditioning in the pipeline during pumping from the mixing station to the trench may lead to a significant reduction in the measured rheological properties. If measurements of the fluid properties of the slurry are to be used for control purposes, it is very important that the tests are undertaken at a defined age on samples of known shear history, e.g. from a pump or homogenising tank and that the time of mixing and the time of sampling and the source of the sample are recorded.

Because of the sensitivity of rheological properties of cement-bentonite slurries to time, shear history etc., their detailed measurement cannot be justified on site, unless it will provide useful information either about the performance of the slurry or for quality control purposes.

2.3.1. Slurry stability, bleeding

The slurry should be stable with minimal segregation, so that solids do not settle out and significant free water (bleed) does not develop at the surface. Any bleeding represents loss of useful volume of slurry. Excessive bleeding suggests instability of the slurry.

Bleed water can drain rapidly from trenches where the near-surface soils are permeable, and this can lead to trench instability if there is substantial bleed. In ground where the surface soils are of low permeability, bleeding is likely to be more apparent as the accumulating water may be slow to drain. Thus the appearance of bleed water should not be automatically assumed to demonstrate a poor slurry. When assessing the significance of bleed water in a trench it will be necessary to consider:

- the amount of bleed that occurs in laboratory tests,
- the drainage conditions at the sides and base of the trench,
- the self weight consolidation stresses acting within the slurry,
- that bleed is a slow process and in deep trenches it may be halted by the set of the slurry before the full bleed capacity has been reached.

2.3.2. Slurry loss to ground, penetration and filtration

Losses to the ground may occur by two principal mechanisms: penetration and filtration. With penetration there is a bulk loss of slurry into voids in the soil. Penetration may be reduced by increasing the gel strength of the slurry. The penetration of the slurry into the soil, L, may be estimated from the following formula:

$$L = 0.15\, DP/\tau$$

where P is the excess hydrostatic pressure of the slurry, τ is the gel strength and D is the effective diameter of the soil particles.

Thus, for a slurry of gel strength 10 Pa in a soil with an effective particle diameter of 10 mm, penetration might be of the order 1.5 m per metre head difference between slurry and groundwater pressures. The constant 0.15 in the equation is a guide figure and will vary with the soil conditions and the slurry. Penetration is unlikely to be significant in soils finer than medium sands. If penetration is a problem it may be necessary to increase the clay concentration or add a blocking material to the slurry.

Filtration will occur as a result of the hydrostatic pressure difference between the slurry in the trench and the groundwater on either side of the trench (this pressure differential is essential for trench stability). Filtration results in the loss of water from the slurry in the trench, with its deposition as solids on the walls of the trench. In permeable ground a cement-bentonite slurry will tend to form a significantly thicker cake than that formed from a bentonite slurry under similar conditions.

Filter loss may be regarded as a comparable parameter to bleed i.e. as bleed under pressure. Filtration will be influenced by the properties of the slurry and the adjacent soil. The filtration properties of a slurry (rate of loss and cake thickness) may be assessed with the American Petroleum Institute fluid loss test rig. The fluid loss of cement-bentonite slurries is normally substantially greater than that of simple bentonite slurries.

High filter loss can lead to trench instability, especially in non-cohesive soils of low to moderate permeability, because the lost fluid will increase local pore pressures if it does not immediately drain into the adjacent soil. In more permeable soils fluid loss is rapidly dissipated and thus should not compromise trench stability (unless it results in a significant drop in the slurry level in the trench). Collapse of slurry trench cut-off walls is rare, but in unfavourable soils or if the groundwater level is high or tidal with high peak levels, the potential for instability should be considered and appropriate measures adopted.

In principle if there are problems with filter loss, e.g. excessive reduction in slurry level in the trench prior to set, filter loss control agents could be added to the slurry. However, there is little experience with the use of these materials in cement-bentonite slurries and proving trials could be necessary (see also the notes on admixture use in Section 3.1.7).

The filter cake that develops on the wall of a slurry trench and slurry that penetrates into the ground will set, providing additional low permeability components to the wall.

2.4. Set properties of slurry

The set properties of cut-off walls have been the subject of much debate, largely because when walls are required for pollution control, their performance may be defined or constrained by Regulatory Authorities. These are less concerned with the nature of the material that is used to form the barrier than with its performance. It should not be assumed that cement-bentonite or cement-bentonite-geomembrane systems are infinitely adaptable and can be modified to meet all regulatory requirements. They are a distinct class of materials with a distinct range of typical properties just as concrete, steel or timber. Those involved with the specification of cut-off walls should recognise these limitations and not require properties of cement-bentonite systems that cannot be achieved—or at least not consistently in the field.

For example, some specifications required the set material to have a permeability of less than 1×10^{-9} m/s coupled with a strain at failure of greater than 5% under low or zero effective confining pressure. A cement-bentonite material with such properties does not currently exist (or if it does it cannot be reliably produced over the full extent of a major wall).

2.4.1. Permeability

Over the last two decades, the permeability specification for cut-off walls has moved from less than 1×10^{-8} m/s, the value required for earth dams, to less than 1×10^{-9} m/s which appears to have been borrowed from landfill liner designs. However, the typical landfill requirement of a 1 m minimum thickness is seldom applied (it can be relatively expensive to form cut-off walls of such thickness). The permeability could be factored to maintain a constant permittivity of less than 1×10^{-9} s^{-1}, but this has not been common practice.

However, specifications have been proposed that require permeabilities of 5×10^{-10} m/s or lower to ensure that the 1×10^{-9} m/s limit is achieved with a reasonable margin for error. The permeability of cement-bentonite materials is sensitive to the age at test and the time under permeation, both factors markedly reducing the permeability. Originally, permeabilities were specified at 28 days, but because of the tighter specification requirements this has generally moved to at least 90 days—or a requirement on the contractor to maintain testing until the required permeability is demonstrated. Specifying substantial ages at test means that on many small-to-medium sized jobs the contractor will have finished the work and demobilised before the first test data are available. If this is unacceptable the specifier may require a 28 day control value not as the design permeability but as a demonstration that the design value can be achieved at a reasonable age (see Section 2.4.5).

2.4.2. Strength

At the present time specified strengths vary very substantially. Specified values have included: greater than 10 kPa, 50 to 300 kPa, less than 500 kPa, greater than 300 kPa etc. and it is clear that there is no consensus. Traditionally, strength has been used as an index by which to assess the quality for engineering materials. Thus, there has been a tendency to specify minimum strengths.

It will be necessary for cut-off designers to address the issue of strength as it relates to their particular application, and to set an appropriate value, subject to checking that it is realistic for a cement-bentonite material. If it is not, an alternative wall type must be selected. A realistic minimum strength for many applications is 50 kPa, and a minimum of 100 kPa is used in the Specification. It must be recognised that, in common with all cement-based materials, the results of strength tests on cement-bentonite materials will show some scatter and that this scatter may be markedly greater than that for concrete.

Consolidation

When considering the behaviour of materials such as cement-bentonite, a topic that ought to be addressed is consolidation of the material under the in-situ stresses. In practice, these stresses are normally safely below the strength of the materials.

Strain at failure

The practice for many years has been to specify a strain at failure of greater than 5% in a consolidated drained triaxial compression test. This figure has become enshrined in repeated specifications, though there are occasional variations, and figures in the range 3 to 7% have also been specified. (A few specifications have used a related parameter, deformation modulus, see below.) The requirement for greater than 5% strain persists even though it has long been realised that it is very difficult to achieve. It is impossible to achieve if failure is defined, as it should be, as the onset of cracking within the material and thus the possible increase in permeability rather than the point at which the stress begins to drop in a compression test, i.e. the point of mechanical failure. A reasonable approximation to the strain at failure criterion was possible when the permeability requirement was less than 1×10^{-8} m/s, but with the current requirements for less than 1×10^{-9} m/s or lower, the strain condition cannot be reliably achieved. A combination of a strain at failure at 5% or more (even if defined as the point of mechanical failure) and permeability of less than 1×10^{-9} m/s or less is unlikely to be achieved throughout a job (if at all) if the effective confining pressure used in the test is required to match that which may occur in the ground. However, 5% strain and greater, prior to mechanical failure, can be achieved if the sample is tested under confined drained triaxial conditions with an effective confining pressure at or greater than about 50% of its unconfined compressive strength. These conditions very seldom pertain in the field, especially at the top of a cut-off wall. When specifying strain at failure tests many specifications have been silent on the question of the required test pressure (presumably deliberately) and thus the contractor was at liberty to use an effective pressure sufficient to ensure a strain at failure of greater 5%.

It appears that a strain at failure criterion is not generally applied in Europe; certainly the typical cut-off mixes used in Europe will not achieve 5% strain at failure. If the 5% condition is removed, the 1×10^{-9} m/s permeability can be achieved more surely by increasing the overall cementitious content. This will reduce the sensitivity of the material to drying and should generally increase its durability however, the materials would be markedly stronger.

The UK situation was clearly nonsensical and in an attempt to resolve the situation the strain at failure criterion has been deliberately omitted from this Specification. Designers should note that in unconfined compression the strain at failure of a cement-bentonite material is likely to be in the range 0.2 to 2%. If higher values are required then designers should either adopt other barrier configurations, e.g. include a geomembrane or commission research

Deformation modulus

to develop a material appropriate to the confining conditions in the trench.

It may be noted that the BSEN 1538:1999 requires that the deformation modulus (that is the stiffness) of the cement-bentonite material be specified by the Engineer to 'satisfy the functional requirements of the wall' but there is no mention of strain at failure. However, the standard does not specify the test conditions for the deformation modulus, age at test, strain level, drainage conditions, confining pressure, strain rate etc.

2.4.3. Physical and chemical durability

Cut-off walls are often required to control the flow of aggressive liquids such as leachates from landfills or contaminated land. These liquids may contain many different chemical species and their composition and concentration could vary not only with position along the wall but also with time. At present, there are almost no design data on the performance even of concrete in such environments. Work is currently in progress at the Building Research Establishment on both the basic behaviour of cut-off materials and their behaviour in contaminated soils. However, it will be some considerable time before a sufficient body of data is available to develop definitive guide rules.

If chemical compatibility testing is to be carried out then it is important that the following are considered:

- **The relative quantities of the test liquid and the cut-off material.** Aggressive materials in the test liquid (e.g. acids) can be depleted by reaction with the cut-off material, and thus the results with small relative volumes of the test liquid may not be representative of the long-term behaviour of the material in-situ. It will be necessary to consider the relative volumes for all types of test including permeation and immersion tests (see also Section 6.2.4.1, relating to reaction on permeation).
- **The test duration.** It is appropriate to note that in-situ tests on the behaviour of concrete in soils containing sulphates are still on-going after 25 years. If data on long-term behaviour of cut-off materials are required, then it may be more appropriate to investigate the performance of existing walls in comparable environments than to wait for laboratory test results.
- **The confining pressure.** The confining pressure that acts on the material in-situ will be different to that applied to samples under test. Unconfined samples when subject to expansive reactions (e.g. sulphate attack) may crack and effectively disintegrate. Under modest confining pressures the cracking can be largely inhibited so that there would be only a modest increase in permeability. Under higher confining pressures, if the chemical attack weakens the material, consolidation can occur to the extent that there is actually a reduction in permeability. Thus, there may be some long term consolidation of cement-bentonite materials in aggressive soils.

It should be emphasised that pure water is aggressive to all cement-based materials, as it can leach lime (calcium hydroxide) from the cementitious phases and ultimately cause some degradation. Curiously permeation with water seems to cause a reduction rather than increase in permeability for cement-bentonite materials. The reasons for this have not been fully quantified but are likely to include:

- Leaching of lime will cause some loss of strength. As a result there may be some consolidation of the leached material and this may contribute to the reduction in permeability.

- Most permeants including distilled water, if it has been allowed to come into contact with the atmosphere, will contain some dissolved carbon dioxide or carbonate species. These will react with the lime in the cement-bentonite to precipitate calcium carbonate, a reaction product which is known to cause pore blocking and reduction in permeability in concrete, and which will be equally effective for cement-bentonite. Indeed if permeants rich in carbon dioxide are used, for example, landfill leachate, there is an immediate reduction in permeability which can be of an order of magnitude or more, though later reactions including the formation of soluble calcium bicarbonate may lead to a subsequent increase in permeability, though the time at which this occurs may be very extended (some estimate of this time can be made by reaction modelling).

The change of permeability with time for water permeation can be estimated from the following formula:

$$k_{(t)} = k_{28}(t/t_{28})^{-n}$$

Where $k_{(t)}$ is the permeability at time t, the exponent n is typically in the range 1 to 2 and thus, for example, there may be a reduction of the order of 3 to 10 times in permeability between 28 and 90 days of permeation. However, the equation must be used with caution as the actual permeability will tend to a limiting value which will be achieved once the full length of the sample is in equilibrium with the permeant. The time to reach this value may be measurable in years. The equation over predicts the reduction in permeability after about a year and indicates an ever decreasing permeability with time.

The effect of leaching by water may be used as a base-line for the estimation of the effects of minor (low concentration) contaminants in a permeant on the performance of a cut-off material.

Limitations

When specifying cut-off properties, it is to be remembered that cement-bentonite slurries have a dual role of supporting the trench during the excavation phase and forming the final barrier, unless a geomembrane is used. Mix design is therefore substantially constrained, and in practice only a limited range of mix proportions are potentially useful. If a high strain at failure or resistance to aggressive chemicals is important, the use of a geomembrane has to be considered as there are generally more data available on the performance of this type of material than for cement-based systems.

There has been a tendency for specifications to place the entire problem of durability on the contractor and typically specifications were of the type '*the wall shall be resistant to ..... chemicals (e.g. leachate) such that its permeability will not exceed 1×10^{-8} m/s over a period of 25/50/100 years*'. The very extended time-scale necessary for permeability/durability tests does not seem to have been generally appreciated. Also when chemical analyses are provided (e.g. for leachate) they do not always include fundamental inorganic parameters such as sulphate, magnesium, ammonium ion, etc., but focus on complex organics at trace levels, the effects of which cannot be quantified on the basis of present knowledge or experience. It should be recognised that a specification which requires the use of a cement-bentonite effectively specifies the durability of the resulting barrier, and that this can be altered only to a modest extent by varying the mix design.

Currently, most cut-off wall mix designs are based on cement-slag-bentonite mixes at high levels of slag replacement (e.g. 60 to 80% slag by weight of total cementitious material) as these give the lowest permeability. However, slag mixes, when unconfined, are very sensitive to sulphate attack in immersion tests. Under confined conditions effects are more limited. Pure Portland cement and Portland cement/pfa mixes are markedly less sensitive under these conditions, but show poorer initial (and perhaps long-term) permeabilities. In this respect cement-bentonite mixes appear distinct from structural concrete where slag replacement generally reduces sensitivity to sulphate attack. This 'anomalous' behaviour of cement-bentonite mixes at high levels of slag replacement is probably due to their markedly higher strength than that of pure Portland cement mixes with the same total cementitious content. Damage from sulphate attack on cementitious materials results from the growth of expansive phases within the materials. In weak materials this growth may not lead to mechanical disruption of the unconfined material.

A further advantage of Portland cement and Portland cement/pfa mixes is that higher cement (or cement plus pfa) contents can be used, whilst maintaining low strength, and thus they are less sensitive to drying and may be better able to survive reactive chemical attack (see also Section 3.1.5).

Design life

When considering design life it should be kept in mind that the first slurry trench cut-off wall in the UK was installed in the early 1970s, though some plastic concrete cut-offs had been installed under earth dams prior to this. Unfortunately, few cut-off walls have been subject to any monitoring in service.

In principle the life of a cut-off wall may be extended if there is always a tendency for inward flow of clean groundwater through the wall rather than a tendency for outward flow of contaminated water. It can, therefore, be advantageous to keep the internal water level slightly below the external one, if this is possible (see also the comments on hydrogeological survey in the Introduction to these Notes for Guidance).

2.4.4. Pre-construction mix trials

Pre-construction mix trials may be needed to assess the effect of spoil and/or contaminated spoil/groundwater on the slurry mix. If these are to be carried out by the contractor, details of the tests and procedures etc. will be included in the Particular Specification. It should be noted that these trials can be expensive and will take a considerable amount of time. Careful consideration of the test programme, test procedures, and possible outcomes therefore will be important.

2.4.5. Timing

Cement-bentonite materials develop their properties (in particular, permeability) rather slowly. Thus, the 28-day tests much used in specifications for concrete may be inappropriate for proving the hardened properties of these materials. The problem thus arises that the cut-off wall works could be finished before the first test results are available.

It is therefore necessary to distinguish between quality control tests during the works and tests to prove the long-term properties of the cut-off material. Quality control tests will be mainly concerned with control of batching accuracy and fluid properties such as specific gravity. However, tests of permeability, strength, etc., at 28 days could be specified provided that prior calibration tests have been undertaken to obtain data on the development of the material properties with time.

Because the materials develop their properties slowly, it will also be necessary to allow sufficient time for the proving of mix designs, especially if unusual properties or combinations of properties are required. For example, if 90 day tests results are required, a period of at least 6 months should be allowed.

2.5. Geomembranes   See Section 3.2 of these Notes for Guidance.

3. Materials

3.1. Slurry components
3.1.1. Water supply rate

Cut-off wall construction requires substantial quantities of water and the supply rate can be the limiting factor in cut-off wall formation. The location of the water supply, if one is designated to be used for the works, and the available flow rate and pressure if a water main is the proposed source as well as any limitations on the supply, should be stated.

Compatibility testing

In general, UK mains water has been found suitable for the preparation of bentonite and cement-bentonite slurries. In contrast, untreated sources such as local lakes and streams may be unacceptable.

Waters containing significant quantities of dissolved salts can inhibit proper dispersion of the bentonite. The levels at which dissolved ions inhibit dispersion vary with the type and source of the bentonite. For UK bentonite, magnesium is often the most sensitive ion and may begin to inhibit dispersion at levels greater than about 50 mg/litre. Calcium may inhibit dispersion at levels greater than 250 mg/litre. However, it is difficult to predict whether a water is suitable from chemistry alone. Therefore if there is any doubt about the water, but particularly if non-mains water is to be used, hydration trials should be undertaken. These may take the form of a full investigation of bleeding, rheology and fluid loss, etc. of bentonite slurries prepared with distilled water and the test water. A simplified procedure could be based on bleed alone as this can be a sensitive parameter. A possible procedure is as follows:

- A slurry consisting of the bentonite and the proposed mix water for the site shall be prepared at a concentration of 5%. If there is any significant bleed (more than 2%) at 24 hours and again at 48 hours, after remixing at 24 hours, then the mix water shall be considered unsuitable for use and an alternative source of water must be sought.
- A slurry of 24 hours hydrated bentonite and the cementitious material to be used shall be prepared and the mix checked for bleed over a period of 24 hours according to Appendix A.4. of the Specification. Mixes that show more than 2% bleed shall be considered unsuitable unless otherwise agreed by the Engineer. However, with cement-bentonite mixes there may be other compatibility problems and thus high bleed in this test should not be taken as conclusive evidence that the water is the problem, unless a matching mix prepared with distilled water does not show the bleed.

3.1.2. Bentonite powder

The bentonite used in the UK for slurry trench cut-off walls is almost always a calcium bentonite converted to the sodium form with sodium carbonate, that is, a sodium-activated bentonite. In Europe and particularly in Germany, calcium bentonite may be used in the cut-off slurry as well as sodium-activated bentonite. Calcium bentonites do not disperse in the same way as sodium bentonites to form thixotropic non-settling suspensions. Thus, by using calcium bentonite, the clay content of a slurry can be increased without it becoming unacceptably thick, but special design procedures are necessary.

At present there is no satisfactory specification for bentonite for slurry trench works. The Oil Companies Materials Association (OCMA) specification DFCP-4, Drilling Fluid Materials Bentonite (or reprints of it, e.g. the Engineering Equipment and Materials Users' Association Publication No 163, Drilling Fluid Materials) are often cited, but involve a relatively narrow spectrum of tests. Most cut-off wall contractors prefer to use civil engineering grade bentonites that have been tailored by the suppliers to meet the perceived requirements of the construction industry, but not necessarily cut-off wall performance. Many civil engineering grade bentonites would not meet the OCMA specification as the OCMA grades can be finer ground and have lower moisture contents.

Although the source of a bentonite may be more easily specified than for a cementitious materials, it is a natural material and there can be variations in the properties of the material from a single source, especially in the longer term or if a new section of a quarry is opened.

3.1.3. and 3.1.4. Portland cement and ground granulated blastfurnace slag

Standard specifications for Portland cement and blastfurnace slag are written around the properties required for concrete; materials such as cement-bentonite slurries are not considered. Nor are issues such as slurry rheology. Cut-off wall contractors therefore have to accept cement and slag produced to specifications which are not focused on their requirements.

It should be noted that both the slag and the cement supplied by the manufacturers to a cut-off site could be from several different works and the source of supply may be outside the control of the contractor. Even a specification that requires the material to come from a single works may not succeed in ensuring that it has been manufactured solely at that works.

3.1.5. Pulverised fuel ash

Pulverised fuel ash can improve the resistance of cut-off materials to chemical attack (it does, for example, reduce the sensitivity to sulphate attack); the unburnt carbon content (inevitably a pfa will contain some unburnt material) may usefully sorb organic materials and so retard their migration through the wall. However, relatively high proportions of pfa (by weight of total cementitious material) are necessary to achieve enhanced chemical resistance. As a certain minimum cement content is necessary to achieve set of the slurry within a reasonable time, pfa-cement-bentonite slurries will have a significantly higher total content of cementitious material than slag-based mixes. In general, cement-bentonite materials containing pfa will have a higher permeability, at least at ages of up to about 1 year, than mixes of comparable strength containing slag as pfa is less reactive than slag in cement-bentonite systems.

3.1.6. Certification

Cut-off walls require substantial quantities of raw materials. Certificates for cementitious materials can cause problems. Under current standards, the same certificate may be supplied with all deliveries of cement, slag, etc. made during a period of one month. This makes it difficult to track the effects of any variations between deliveries. If individual certificates are required for every delivery this should be specified to the supplier and an allowance made for the cost of the additional testing.

3.1.7. Admixtures

In the past, the use of admixtures has been inhibited by specifications of the following type '*the contractor must demonstrate that any additive used in the slurry will not impair the long-term performance of the cut-off wall'*. This is a reasonable concept, but a condition practically impossible to demonstrate. Admixtures became more widely used when geomembranes became a frequent part of cut-off walls as the problems of geomembrane installation, particularly after a weekend break, necessitated extended set times. Admixtures are now more commonly used and are often based on proprietary admixtures for concrete. Admixtures may be useful not only in controlling set time but also in improving the fluidity of a mix.

The timing of the addition of any admixture should be consistent for all mixes as the behaviour of an admixture may be different if, for example, it is added to the bentonite before the cement addition or concurrently with the cement.

3.2. Geomembrane

Before considering the properties of the geomembrane some comments on the cement-bentonite-geomembrane composite cut-offs are appropriate:

The geomembrane panel

A geomembrane panel may consist of three elements:

- the geomembrane sheet
- an interlock to join geomembrane panels
- a weld which joins the interlock to the panel (the welded joint or seam)

If performance or quality assurance tests on the interlock are required these must be detailed in the Particular Specification.

Permeability

For a composite wall it will be necessary to consider the overall performance of the geomembrane-slurry-interlock system, the manner in which the composite action is achieved and how this might be compromised. In particular it will be necessary to consider the permeability of the interlock system to the fluids to be retained (aqueous, non-aqueous, and gaseous), the potential effects of any chemical/physical interaction, and the possibility of seasonal groundwater level/moisture content changes in the ground.

It should be noted that an intact HDPE geomembrane may be effectively impermeable to inorganic species such as sodium and chloride ions. However, organic species including materials such as chlorinated solvents may be able to diffuse through a geomembrane comparatively rapidly.

Chemical compatibility

Consideration of chemical compatibility should include:

- the performance of the geomembrane and the geomembrane jointing system both as regards installation of the panels and long-term performance as a liquid and/or gas barrier,
- the contaminants that may be present in the ground,
- the chemical conditions imposed by the cement-bentonite (e.g. the high pH (13), or dissolved salts in the pore water could be an issue for some jointing systems).

The geomembrane and jointing system should offer satisfactory control of the fluids to be contained under the chemical and physical conditions that prevail in the trench as influenced by the adjacent ground. It also has to be mechanically robust, so that it will not be damaged by the installation process or by stresses applied during the installation or manipulation of subsequent geomembrane panels.

3.2.1. Geomembrane material

To date most specifications for HDPE geomembranes in cut-off walls appear to have been developed from specifications for landfill liners. It is recognised that in due course a geomembrane specification particular to the conditions in a cut-off wall may be developed. However, in the absence of such a specification it is necessary to borrow from landfill specifications and to add additional tests where necessary.

The approach to geomembrane materials is to use a standard specification and, in the absence of a British Standard (a draft British Standard for the classification of geomembranes is being produced by BSI Sub-Committee B/546/8), a European CEN document or an international standard, the US National Sanitary Foundation (NSF) standard has been adopted. However, for cut-off walls some additional requirements, over and above the NSF standard, are necessary and these are detailed in Table 1 of the Specification.

The NSF standard

The requirements of the NSF standard relate to the geomembrane's physical and chemical properties determined by the following tests:

- thickness (nominal thickness and lowest individual reading)
- specific gravity (minimum requirements)
- tensile properties (tensile strength at yield and break, together with elongation at yield and break)
- tear resistance (minimum requirements)
- low temperature impact (maximum allowable failure temperature)
- dimensional stability (maximum allowable strain)
- environmental stress cracking (bent strip test, minimum number of hours with no failures)
- puncture resistance (minimum requirements)
- carbon black content (allowable range given)
- carbon black dispersion (acceptable levels given)

In addition, there are bonded seam strength and seam peel adhesion requirements for factory seaming.

Particular attention should be given to the chemical compatibility of the geomembrane, both in terms of the chemical composition of the slurry wall and the composition of the contamination to be contained. It is the duty of the designer, on a site-specific basis, to assess the ground conditions to be encountered and to state the compatibility requirements in the Particular Specification and how they are to be established (i.e. test procedures etc.).

Compatibility tests such as the US Environmental Protection Agency Method 9090 can be used to determine the effects of chemicals on geomembranes, and will assist in deciding whether a given geomembrane material is acceptable for the intended application.

Geomembrane properties
Physical properties

The physical properties are important for the identification of the geomembrane. Measurement of the density will show whether the geomembrane has been manufactured from high density polyethylene. The determination of the density of a geomembrane is carried out using a density column, which comprises a long glass column containing liquid varying from high density at the bottom to low density at the top. Materials of known density are immersed to form a calibration curve, and small pieces of a geomembrane are then placed in the column, allowed to settle to an equilibrium level, and the density then read off.

Variations in the thickness of the geomembrane could suggest problems associated with the manufacturing process. The thickness of a geomembrane is determined by weighing a sample of known size, measuring its density, and calculating the average thickness.

Chemical properties

Conceptually, the chemical ageing process of a HDPE geomembrane can be considered in three distinct stages; depletion of antioxidants, induction time to the onset of polymer degradation and degradation of the polymer. The purpose of antioxidants in a HDPE geomembrane formulation is to prevent degradation during processing and to prevent oxidation reactions taking place during the first stage of service life.

The Oxidation Induction Time (OIT) is the time required for the geomembrane test specimen to be oxidised under a specific pressure and temperature. The length of the OIT indicates the amount of antioxidants present in the test specimen.

Strength properties

Strength properties are required for design purposes, and also to check that the material will not be damaged during installation and handling. The wide-width tensile behaviour of a geomembrane is considered to be more representative of the stress conditions likely to be encountered on site than the more common 'dumb-bell' test. In the wide-width test, a 200 mm wide specimen is gripped across its entire width and a uniaxial load is applied until the specimen ruptures. The dumb-bell test, however, is the industry standard and is a useful quality control test.

The tear resistance test uses a template to form a test specimen shaped with a 90° notch where a tear can initiate. The two ends are pulled apart in a tensile testing machine and tearing proceeds across the specimen perpendicular to the application of the load.

The Federal Test Method Standard 101C method 2065, for measuring puncture resistance, requires measurement of the force necessary to pierce a geomembrane sample with a 3.1 mm diameter probe.

Endurance properties

The geomembrane may suffer from ultraviolet degradation if exposed to sunlight for any prolonged period. Protection is afforded to the geomembrane by the addition of carbon black to the base polymer. This carbon black content is measured gravimetrically after pyrolysis of the sample under nitrogen.

Stress cracking is associated with all semicrystalline polymers, but in practice the potential becomes more developed as the polymer becomes more crystalline. Environmental stress cracking (ESC) is stress cracking associated with an aggressive environment, and high density polyethylene has been shown to suffer from ESC. The test for assessing the performance of a geomembrane comprises subjecting a dumb-bell shaped notched specimen from a geomembrane sheet to a constant tensile load in the presence of a surface active agent at an elevated temperature. The time to failure of the test specimen is recorded. The results of a series of such tests conducted at different stress levels are presented as a plot of stress level against time to failure. This generates the entire applied stress versus time to failure curve. In this test, it can take well over 1000 hours for the establishment of the brittle region of the curve. The concept of the single point NCTL (Notched Constant Tensile Load) test is to select a stress level near the initial portion of the brittle region of an applied stress–time to failure curve. The failure time of the sample should then exceed a specified value.

The degree of carbon black dispersion is considered to give an indication of the effectiveness of the ultraviolet screening of the carbon black. An examination of the dispersion may also identify any foreign matter, spots of unpigmented polymer resin or resin degradation. Microscopical examination is carried out and the observations made are compared against observational standards.

Temperature

Prior to installation, the geomembrane could be subjected to temperatures varying from below 0 to 70°C. The upper temperature specified is due to solar gain with the geomembrane in direct sunlight on the surface. When installed within a slurry wall the maximum temperature could be as high as 40°C resulting from the hydration of the cement. In the long term the ground temperature is likely to be in the region of 12°C in the UK. If the geomembrane is to be used at temperatures outside the range recommended by the manufacturers, their guidance must be sought.

3.2.2. Manufacturer's quality control

Manufacturer's quality control within the geomembrane industry is generally of a high standard, but the items required on the quality control certificates are a minimum. As stated in the Specification the testing methods specified are suggested methods, since the Specification cannot dictate to the geomembrane manufacturer which tests to use for the quality control. In particular German DIN standards may be comparable and therefore acceptable.

3.2.3. Delivery, handling and storage

The geomembrane panel fabrication may be carried out on the job site, but it is more likely to be at a location away from site.

3.2.4. Contractor's experience/ manufacturer's experience

The experience requirement of the installation contractors has not been quantified. Such experience is important and the onus is on the Contractor to demonstrate competence.

3.2.5. Welded joints

The quality of workmanship required for the welded joints is to be of comparable standard to that for geomembranes in landfill liner applications, where full quality assurance measures are in place. It is necessary to distinguish between welded joints and interlocking joints. The term 'seam' is used to describe the welded joint that is required to join an extruded interlock element to the geomembrane.

It is strongly recommended that the Client engages the services of a specialist geomembrane quality assurance monitor. This role could be undertaken by the Engineer for the works, or preferably by an independent third party. A quality assurance plan should be drafted by the quality assurance monitor and submitted to the Client, and if required to the regulatory authority, for approval prior to work commencing on site.

The procedures regarding welded joints are again a minimum, and the Engineer, together with the quality assurance monitor, should highlight any specific requirements in the Particular Specification.

The welded joint should be prepared under factory conditions or such conditions as are specified by the manufacturer. Any welding should be undertaken only under conditions specified by the manufacturer.

3.2.6. Interlocking joints

Interlocking joints are typically proprietary systems developed by geomembrane manufacturers. Most interlocks are formed by welding a pre-formed interlock element to the edges of the sheet. Quality assurance procedures, for example tensile tests, on the interlock and this welding therefore may be required in addition to those for the geomembrane and the interlock element itself.

4. Slurry production and quality control

4.1. Slurry preparation plant

The necessary time of hydration for the bentonite will depend on the type of mixer used and the type/source of the bentonite. If times shorter than 8 hours are to be used, they should be demonstrated to be satisfactory by data from mixes prepared with the proposed site materials and bentonite and cement-bentonite mixing plant.

4.2. Batching

4.2.1. Accuracy

Experience has shown that quality control checks based on the rheological properties of a cement-bentonite slurry are of little value as the results are strongly influenced by the age of the slurry at the time of test and the elapsed time from when it was last mixed, pumped, sheared, etc. The most effective method of controlling the slurry is to ensure that it is accurately batched and uniformly mixed.

A batching accuracy of ±3% is required for cement-slag mixes at slag replacement levels of over about 60%. The properties of these mixes are sensitive to the slag/cement ratio, thus tight mixing tolerances are necessary. Mixes based on pure cement or cement/pfa are less sensitive to the mix proportions and thus the batching accuracy need not be so rigorous and a greater tolerance may be permitted.

4.2.2. Calibration

The batch quantities of cement, bentonite and other materials may be small as compared with those used in concrete batching. It is therefore important that load cells and dispensing equipment are appropriate to the quantities to be batched. The accuracy of dispensing should be checked at the start of any cut-off wall contract and at regular intervals thereafter. Whenever possible this checking should be carried out directly by weighing the amount of material dispensed, rather than indirectly by checking load cell calibrations with weights.

The quantity of water dispensed should be checked by measurement of the dimensions of the mixing tank and the depth of water, making due allowance for the volume of any pipework included in the water-filled circuit or with an in-line flow gauge graduated and calibrated to suitable precision.

4.3. Quality control

As noted in Section 4.2.1, quality control based on slurry rheology (determined by a direct reading viscometer or Marsh funnel, etc.) of the mixed cement-bentonite slurry is of limited value. Tests on the bentonite slurry alone may help in identifying gross errors in mixing. However, if they are to be of any comparative value, they must be made on slurries of the same age and immediate shear history. Thus samples taken from the bentonite delivery pump should give comparable results if of comparable age from mixing.

Batching accuracy can be confirmed by measurement of the moisture content of the bentonite and cement-bentonite slurries. Allowance must be made for the moisture content of the bentonite as supplied; typical values can be in the range 10 to 22% (The moisture content is defined as weight of water by weight of dry solids. In much of the trade literature on bentonite, the moisture content is quoted by total weight, i.e. weight of water plus dry solids). For cement-bentonite slurries it also may be necessary to make allowance for non-evaporable water combined in the cement hydrates. This will be age dependent and may be small for young slurries.

However, moisture content measurements will take some hours to complete and therefore are not convenient for immediate on-site control of batching/mixing. This can be achieved by density measurement but the industry standard test procedure, the American Petroleum Institute (API) mud balance, is not of sufficient resolution to be of use. It is readable to at best 0.005 g/ml which corresponds to a batching accuracy of order 15 kg/m^3; as the total bentonite content might be of order 45 kg/m^3 and the total cementitious content 125 kg/m^3, this is clearly inadequate. More accurate measurements can be made with a fixed volume container (a volume of order 1 litre may be appropriate) and an electronic balance though even with such equipment it will be difficult to achieve a repeatability of better than 0.002 g/ml. A procedure for the measurement of the specific gravity of a slurry (i.e. its density relative to that of water) is given in Section A.3 of Appendix A of the Specification.

There can be problems of air entrainment affecting the specific gravity. Bentonite and cement-bentonite slurries are gelling materials and so can easily trap air bubbles. Air may be entrained during the mixing process and will also be present in the pore space, and on the surface of the dry bentonite and cement powders prior to mixing. Thus, there always will be some air in a slurry after mixing.

4.3.1. Bentonite suspension

4.3.1.1. Sampling

As discussed in Section 2.3, the properties of both bentonite and cement-bentonite slurries are complex, being time and shear history dependent. Rheological properties of bentonite suspensions will be sensitive to the time elapsed since mixing, and the shear history.

4.3.1.2. Viscosity

The measurement of slurry viscosity can be used as a quick check that a reasonable degree of hydration has been achieved in a bentonite slurry if it is specified that a minimum Marsh funnel flow time is achieved. As many different flow funnels are used in the construction industry, it is appropriate to note that the Marsh funnel has a discharge orifice diameter of 3/16 inch (4.76 mm), and is to be filled with 1500 ml of test slurry and the discharge time for 946 ml (1 US quart) or 1000 ml is to be recorded. For water, the Marsh funnel flow time for 946 ml is 26±0.5 seconds, and for 1000 ml it is 28±0.5 seconds. The test procedure is set out in Appendix A of the Specification.

4.3.1.3. Specific gravity

Specific gravity can be used as a check on batching accuracy, but as noted in Section 4.3, considerable precision is required. The API mud balance is not suitable for measuring/controlling the bentonite or cement content of cement-bentonite slurries.

4.3.2. Cementitious slurry

4.3.2.1. Sampling

Samples of the cement-bentonite slurry may be taken from the mixer, the delivery pipe or the trench. Samples from these different locations are likely to be of different ages when tested and certainly will be of different shear history and hence they will show markedly different rheological properties. This is a feature of the slurries and should not be taken as an indication of poor quality control of materials, mixing or batching.

4.3.2.2. Specific Gravity

As noted, the API mud balance is not suitable for checking the materials content of cement-bentonite slurries. However, the instrument may be useful for measuring the specific gravity of slurry samples taken from the trench when a main concern could be spoil contamination when slurry specific gravities may be markedly above that of the freshly mixed slurry.

Viscosity

Many cement-bentonite slurries will be too thick to give reasonable test times (less than 60 seconds) with the Marsh funnel. Flow may stop completely before the required quantity of slurry has been discharged. If measurements of the fluid properties are required for such slurries a direct reading viscometer or a flow trough may be used, but note the comments in Section 2.3 on the limited utility of rheological data. However, tests with this and other rheological test equipment may be of considerable use in the laboratory when developing new slurry systems or in the field when evaluating such systems. A more detailed discussion on the measurement of the rheology of cement-bentonite slurries is given in Section 2.3. For the reasons set out above and in Section 2.3, no viscosity tests are included in the Specification for the cement-bentonite slurry.

4.3.2.3. Bleed

See Section 2.3.1 of these Guidance Notes for discussion and Appendix A of the Specification at Section 4 for the test procedure.

4.4. Temperature

See Section 5.1.7 of these Notes for Guidance.

5. Slurry wall construction

5.1. Slurry wall construction

5.1.1. Excavation

Currently, in favourable ground conditions, cut-off walls can be excavated to a depth of approximately 16 m, as a semi-continuous operation with a backhoe or similar equipment. However, recently depths approaching 30 m have been excavated with a backhoe in the UK. More usually for deeper walls, or unfavourable grounds (e.g. those containing hard strata, especially at depths which may prove difficult/impossible to excavate with a backhoe) it will be necessary to use a cable-hung or a kelly-mounted grab, either as the sole method of excavation or to deepen a trench partially excavated with a backhoe.

If long excavation times are expected (hard ground or deep walls) or problems are anticipated or encountered with the excessive suspension of spoil in a cement-bentonite slurry, it may be necessary to excavate the wall in panels under a bentonite slurry, and then to replace this slurry with a cut-off mix. In other words, it may be necessary to employ a two-phase wall procedure rather than a single-phase procedure. For two-phase walls, the cut-off backfill mix must be significantly denser than the excavation slurry. For example, the German Geotechnical Society publication *Geotechnics of Landfill, Design and Remedial Works, Technical Recommendations — GLR*, 1993, prepared by the European Technical Committee 8 of the International Society for Soil Mechanics and Foundation Engineering, recommends a maximum density of 1.3 Mg/m^3 for the excavation slurry and a minimum density for the backfill of 1.8 Mg/m^3. As a result, it is normal practice to use a cement-bentonite-aggregate, plastic concrete backfill, although a minimum density of 1.8 Mg/m^3 may not be sufficient to ensure a stable non-settling plastic concrete. For deep walls, such materials might also be required for their greater strength and to prevent previously placed material from slumping when adjacent panels are excavated.

Walls may be excavated with a backhoe as a semi-continuous operation. However, if a cable-hung or kelly-mounted grab is used as the sole method of excavation, it may be necessary to excavate the wall in panels on a hit-and-miss basis (i.e. excavating panels 1, 3, 5, etc. followed by 2, 4, 6, etc.). The use of a grab will not only require additional craneage but could also significantly reduce the excavation rate.

5.1.1.1. Excavation depth

If there is uncertainty about what the final depths will be along a wall, or if the expected depth range would imply changes in the working method or plant and/or the excavatability of the ground may vary, this should be prepared for by establishing separately priced rates for different types of excavation plant. A radical change in programme and plant requirements may follow if it is necessary to change the planned wall excavation procedure during a contract.

5.1.1.2. Guide walls

If the wall is to be excavated in panels with a cable-hung or kelly-mounted grab, it may be necessary to use guide walls to provide alignment and to increase the near-surface trench stability. They might also be necessary for certain geomembrane walls. If guide walls are used, they should be designed as for structural diaphragm walls. Reinforcement should be continuous between adjacent sections of guide wall and they should be propped apart when not subject to excavation.

5.1.1.3. Slurry level

During the excavation phase, the slurry level cannot in practice be maintained closer to the ground surface than about 0.3 to 0.5 m. If this level is insufficient to maintain trench stability, it may be necessary to lower the groundwater level by dewatering or to raise the working platform level. See also Section 1.7.1.

5.1.1.4. Ground and toe levels

See Section 2.1 of these Notes for Guidance.

5.1.2. Wall location and dimensions

5.1.2.1. Setting out

Setting out tolerances should be realistic having regard to the planned method of excavation. If guide walls are not used it is likely that there will be over-break at or near the surface so that this part of the excavation will be wider than the nominal dimension.

5.1.2.2. Width

Typically cut-off walls are specified to have a minimum width of 600 mm. However, for the deeper walls excavated with a backhoe the actual minimum width may be set by the dimensions of the excavator arm and could have to be 750 mm or more.

The flow-rate through a wall is proportional to the permeability of the wall and inversely proportional to its thickness (see comments on permittivity in the Introduction to these Notes for Guidance). Thus, thickening a wall will improve its performance as a barrier, but the benefit is typically small. In the UK, the maximum thickness cut-off wall using cement-bentonite walls that has been constructed is of the order of 2 m, although even walls of thickness 1 metre or more are very rare. There is no fundamental reason why thicker walls should not be used, although greater benefit probably could be achieved by using a double wall. In the USA, soil-bentonite walls are often 1.8 m thick.

5.1.2.3. Verticality

Verticality is not a major issue for a wall excavated with a backhoe, because continuity of the wall can be demonstrated by passing the arm/bucket of the excavator along each newly excavated length of wall. However, there may be concern about the effect of verticality on the continuity of the wall through day joints. If the start of new day's work deviates from the old, the overlap between sections may be less than the full thickness of the wall. Verticality also may be important if a geomembrane is to be incorporated in the wall as deviations in vertical alignment between adjacent geomembrane panels may make it difficult to assemble the interlocks.

If a tight tolerance for verticality of a wall excavated with a backhoe is to be set, consideration should be given as to how it is to be measured.

Verticality is of much greater importance for deep walls excavated in panels. For these walls the verticality tolerance must be related to the minimum acceptable overlap between adjacent panels.

5.1.2.4. Toe levels, depths

See Section 2.1 of these Notes for Guidance.

5.1.3. Day joints

To achieve continuity between sections of wall it is necessary to excavate back into previously placed slurry to a distance of typically 0.5 m. This cut-back has to be throughout the depth of the wall and across the full width of the previously formed wall. Verticality at day joints is therefore important but may be difficult to measure. The operator of the excavation plant should be instructed regarding these requirements and the formation of day joints should be supervised and controlled until a satisfactory working practice has been confirmed. The cut-back may be proved through the full depth of the trench by plumbing.

5.1.4. Temporary protection

5.1.4.1. Safety

The area adjacent to a slurry wall and that used for spoil deposit will be very slippery as a result of spilled slurry. This slurry will remain very slippery even when hardened. Appropriate fencing therefore should be provided and the trench covered as necessary, e.g. with frames of reinforcement mesh placed over the trench.

5.1.4.2. Reduction of surface cracking

Temporary capping material, such as plastic sheeting or wet straw, should be deployed at the end of each working day to limit drying and cracking of the cement-bentonite material before installation of the final capping. Covering with sacking is unlikely to prevent cracking and even plastic sheeting is unlikely to be effective in hot weather.

5.1.5. Actions on loss of slurry or collapse of trench

Slurry losses will be highest in permeable ground and especially for walls toed into permeable strata. It has not been normal practice to use fluid loss control agents in cut-off wall slurries, so there is little experience of their use (see notes on admixtures, Section 3.1.7). As cut-offs are usually installed in permeable ground some reduction in slurry level must be expected after the end of excavation and before set (and indeed it may be advantageous as it will build a filter cake, see Section 2.3.2, filter loss). Preferably, allowance should be made, in the design, for the drop in slurry level so that the final trimmed level of the wall is below the settled level of the slurry. Settlement can be made up by the addition of slurry, but there is a risk of a cold joint.

Collapse of slurry trenches is rare but may be associated with high groundwater levels or excessive loss of slurry by bleeding or filtration. Also, some grounds such as non-cohesive silts (including lenses of such material) may be more prone to collapse than cohesive or free draining materials.

If a collapse occurs generally it will be necessary to allow the slurry to harden before undertaking remedial actions which could include local dewatering or change of slurry mix design.

5.1.6. Excessive specific gravity of slurry in trench

During excavation of the trench some spoil will always be incorporated in the slurry. However, in some types of ground, and especially fine non-cohesive soils the slurry becomes heavily loaded with spoil particles that are either suspended in the slurry or sink so slowly to the base of the trench that it is impracticable to remove them by excavation. The spoil loading may lead to doubts about the performance of the set slurry. Performance of contaminated slurries can be assessed by pre-construction trials, but such tests would not be undertaken by a contractor unless specified by the Engineer in the

Particular Specification. If such spoil suspension is deemed excessive, it may be necessary to:

- Carry out laboratory trials to confirm that the spoil laden slurry meets the specified properties for the cut-off material. These tests should be carried out along with other design trials prior to the works and should test the slurry up to the maximum credible level of spoil incorporation, including its associated soil moisture.
- Specify a deeper excavation so that excessively dense slurry is restricted to a region below the design toe level of the trench so that the uncertainties associated with the properties of the spoil laden slurry can be limited. The level at which the specific gravity becomes excessive can be assessed by regular sampling and measurement of the slurry in the trench.
- Incorporate a geomembrane in the wall.
- Use a two-phase wall and clean the excavation slurry before placement of the cut-off mix.

In principle the spoil can be removed from a cement-bentonite slurry with a cleaning plant. Indeed, coarse spoil contamination has been removed from a slurry by recirculation through vibrating screens to enable single-phase wall construction with a cutter-miller type machine (hydromill). If cleaning plant is to be employed it is most important that the slurry is designed for use with the plant. For example, the use of equipment such as dewatering screens and/or hydrocyclones to remove fine spoil may also remove some of the cement, slag or pfa from the slurry. Also, spoil removal, unless essential to the excavation process, may not be always beneficial. A heavily spoil-contaminated slurry may behave rather as a plastic concrete (see Section 5.1.1) with the spoil acting as impermeable 'plums' in a slurry matrix. However, this matrix may be slightly more dilute than the original (as-mixed) slurry as some of the water originally associated with the spoil (i.e. the pore water in saturated or partially saturated soils) will be dispersed into the slurry. This dilution will increase the permeability of the matrix but the 'plums' will tend to reduce it if they are left in the slurry and there are no preferential flow paths at the solid-matrix interface. These are complex issues and full-scale trials might be necessary. Cleaning plant will not be normally available on a cut-off wall site and its deployment will be expensive, not only because of the costs of the plant, but also because of the impacts on production. If slurry cleaning is likely to be necessary, rates should be requested at the time of tender.

Because of the incorporation of spoil and possibly chemical contaminants from the ground in the slurry, it is normal practice to take samples of the slurry for control purposes from the trench and not solely from the mixing plant (see Sections 4.3.2.1 and 6.1.3). However, as the effects of spoil or chemical contamination are difficult to predict in advance of the works, if no site-specific testing is included in the Particular Specification, it may be useful to take samples from the mixing plant and/or the delivery point at the trench as references for the properties of the uncontaminated materials. For difficult grounds the contractor may propose to use such samples as the 'contract' samples though this will require modifications to the Specification.

5.1.7. Ambient temperature

Bentonite and cement-bentonite slurries will freeze at temperatures a little below 0°C. Storage tanks and pipe-work therefore should be protected in cold weather.

Frost is unlikely to penetrate to a significant depth into the cut-off wall, especially as the heat of hydration from the cement in a slurry will raise the temperature. Any frost damaged material will be near the surface and should be removed during the capping operation.

5.1.8. Capping

The cap may have several functions including:

- to prevent drying of the cement-bentonite material,
- to extend the function of the underground cut-off wall to above the final trim level of the wall,
- to provide some effective confining stress to the cement-bentonite at the top of the wall and so improve its stress-strain behaviour and its performance if subjected to chemical attack.

These functions are described in Sections 2.2 and 5.1.4.2.

5.1.8.1. Trimming of top of slurry

Any damaged, dried or otherwise unsatisfactory cut-off material should be trimmed from the top of the wall before capping. Capping should follow immediately after trimming to limit drying.

5.1.8.2. Capping detail

The required capping detail will depend on the required function(s) of the cap. The interface between the cap material and the cut-off material is potentially a plane of weakness and should be carefully considered, especially if the interface is to be subjected to fluid pressure.

5.2. Additional measures for geomembrane walls

5.2.1. Installation of the geomembrane

For geomembrane walls, it is necessary to plan to deal with corners or deviations in the line of the wall. Geomembrane can be placed by mounting on a frame and lowering this frame into the trench. Use of an angled frame is unlikely to be practicable and thus if a corner/deviation in the wall is so sharp that the frame cannot be moved around it, the last panel before the corner/deviation has to finish at the corner/deviation. It is not practicable to establish the position of any geomembrane panel joint at the outset of the works as the geomembrane may take up a meandering path in the trench. It therefore may be necessary to 'concertina' one or more of the earlier panels so that the joint at the corner/deviation is at a convenient location. Alternatively if site conditions permit, the location of the corner or deviation may be moved to suit the actual panel position.

5.2.2. Dimensions

The dimensions of the panel will be influenced by the proposed installation procedure. Large panels may be difficult to install in windy weather.

5.2.3. Base levels, bottom details and sloping strata

On early geomembrane walls, it was the practice to mount the geomembrane on a metal frame and hammer the whole assembly into the soil at the foot of the trench so as to improve the base seal. However, because of the risk of damage to the interlocks and associated welding between panels, this is no longer common practice. In general, the geomembrane is not mechanically sealed to the underlying stratum, but the trench is excavated to achieve a specified minimum penetration into the chosen base layer and the geomembrane is then suspended so that it just reaches the base of the trench. In this way penetration of the geomembrane beneath the base layer and a certain minimum flow path length under the geomembrane are ensured.

It is not practicable to cut the base of a geomembrane to match the profile of the base of a trench if it is sloping. If the proposed base alignment is not horizontal, the trench must be stepped so that rectangular sheets can be installed. The steps must be excavated to achieve the specified minimum depth of penetration into the base layer.

5.2.4. Positional tolerance

Some specifications call for the geomembrane to be installed in the centre of the trench, but often no reasons are given for this requirement. While it is possible for the average position of the geomembrane to be at or near the centre of the trench, it ought to be accepted that 'wrinkles' are inevitable and the geomembrane will take up a meandering path both horizontally along the trench and possibly vertically down through it.

5.2.5. Changes in ground level

On sloping ground or if there are sharp changes in level, special procedures will have to be adopted to prevent slurry loss. See Section 1.7.1.

5.2.6. Repairs to damaged sections and joints

Repairs to damaged sections of a geomembrane or essential on-site joints (e.g. to carry a geomembrane around services) must be made in accordance with procedures specified by the geomembrane manufacturer and be subject to appropriate quality control measures.

5.2.7. Temporary stop ends

Temporary stop ends may be necessary to prevent the slurry setting in the interlocks of a geomembrane prior to insertion of the following panel. They also may be necessary during excavation for the following panel to reduce the risk of damaging the geomembrane or the joint.

5.2.8. Temporary fixing and protection of geomembrane

The Specification requires that temporary fixing and protection of the geomembrane shall be carried out by the slurry wall Contractor for a period of seven days and thereafter the responsibility for providing protection measures will rest with the Contractor carrying out the wall capping works. It is important that the geomembrane is protected at all times. Repeated flexing of any exposed geomembrane at the top of a wall, for example by wind action, may cause damage to some types of interlock.

6. Compliance testing for material properties

As already noted, cut-off wall materials may develop their properties rather slowly and thus it may be necessary to carry out proving tests at or after 90 days. On the other hand, it is often desirable and it may be necessary to have some earlier indication of material performance, e.g. tests at 28 days, and indeed for some mixes the material may have hardened sufficiently for tests to be possible at 14 days. However, unless the test facilities were available on site, tests at this age would require modification of the Specification requirement, at Clause 6.1.4, that samples of the cement-bentonite material shall not be transported until age 14 days.

6.1. Sampling for testing for set properties
Sampling equipment

As walls may be of considerable depth, it is important that the sampling equipment should be capable of functioning under substantial hydrostatic pressure and should not leak and so let in slurry until opened at the required depth. It may be appropriate to prove the tightness of the sampling equipment by lowering it to the maximum depth in a trench without operating the slurry entry mechanism. If slurry enters the sampling chamber during such a test then repairs/modification of the equipment may be necessary.

To date, no coring procedure has been developed which has enabled undisturbed samples to be taken from hardened slurry in a cut-off wall. The degree of damage or disturbance depends critically on the amount and type of ground mixed into the slurry. The presence of gravels, in particular, results in such poor quality cores that permeability results are meaningless. If the wall is to be cored, the damage inflicted by the coring operation, not only on the core, but also on the in-situ material should be considered.

Great care also should be exercised when coring in or adjacent to a wall to ensure that the wall is not hydro-fractured by any fluid pressures employed.

6.1.1. Sample tubes

Samples of the slurry for testing can be cast into cylindrical moulds (tubes). During filling, the tubes should be held slightly off vertical and the slurry slowly poured down the side so that air bubbles can escape. At regular intervals pouring should be stopped and the mould tapped to release bubbles. After filling the tubes should be capped and stored upright. Thereafter they should be stored under water in a curing tank until required for testing. It is most important that tubes are watertight. If there is any leakage, it is possible that only water will be lost so that the solids become more concentrated. Also, leakage may cause cracks to develop within the sample.

Cement-bentonite slurries are strongly alkaline and will react with aluminium sample tubes to generate hydrogen gas. **Aluminium sample tubes must not be used since they will prevent the slurry from setting.** It is therefore important to ensure that sample tubes are inert to the slurry. Plastic tubes have been found to be satisfactory.

The tubes should have a uniform smooth internal bore. Any roughness or distortion may lead to substantial damage during extrusion. A recommendation to lightly wipe sampling tubes with mould release oil before filling with slurry to ease demoulding has been questioned. Some recent work suggests that oils may damage the surface of the slurry leading to the potential for damage during extrusion and a more permeable surface layer which may negate the results of permeability tests. It is therefore recommended that sample tubes should not be oiled. Where damage does occur during extrusion, the extrusion pressures can be reduced by reducing the length of the slurry sample. This can be done at the time of casting the sample or after it has set. Reducing the length of slurry to about 300 mm will allow a reasonable amount of slurry to be trimmed from either end of the sample when preparing a 200 mm long test specimen. A tube length of 300 mm may require more samples to be taken from the trench as it is the practice to use 450 mm long samples so that two samples may be prepared from a single tube.

The end caps should be carefully sealed to the tube so that they are watertight. Although it is current practice to used wax to seal the top and bottom of samples, damage is possible during extrusion of the sample if any wax remains adhering to the walls of the tube. Preferably the slurry surface should not be covered with wax.

Sub-sampling from U100s, etc.

In general samples should be tested as cast. Sub-sampling in the sense of coring smaller samples from a larger sample (e.g. 3 by 38 mm diameter sub-samples from a 100 mm diameter sample) will cause unacceptable damage. However, with care, 100 mm diamater samples can be cut from block samples of the material using a soil lathe or similar equipment.

6.1.2. Frequency

The Specification requires that one sample set shall be taken for each 200 m^2 of projected wall area subject to a minimum of one set per day's production. However, it may be noted that the GLR (1993) recommends only one sample for testing of hardened properties from the top and bottom of the wall per 1000 m^2. This may be sufficient to prove the performance of the wall but it will not be sufficient to localise any problems. A sampling frequency of one set per 200 m^2 will produce a substantial number of samples from a typical wall. It is not expected that all these samples will be tested, and this is specifically recognised in Clause 6.2.2 of the Specification which requires that: *'Unless otherwise stated in the Particular Specification, the minimum testing requirement shall be two strength and permeability tests for each 1000 m^2 of wall.'* The frequency of sampling required by Clause 6.1.2 of the Specification means that if problems are identified in any tests, there will be sufficient samples to localise the problem to a reasonably short section of wall.

6.1.3. Location

The Specification requires that samples should be taken from one metre below the top of the trench, and one metre above the bottom. In some earlier specifications there has been a requirement for an additional set of samples from the middle of the trench but, in such specifications, it was normal for only one sample to be taken at each location and not a set. The present form of the Specification is intended to strike a reasonable balance between the number of samples taken and the number of locations sampled.

6.1.4. Care of samples

Cement-bentonite materials are sensitive to disturbance and should be treated with great care. Samples should be carefully packed to avoid damage in transit from site to the test laboratory. The requirement that samples shall not be moved from site until age 14 days is to avoid damage to the material while it is still soft (see also notes on Section 6.2.2).

6.1.5. Storage and transportation

Samples of cement-bentonite will be of no value for testing if they are allowed to dry or to freeze. Samples should be stored at a controlled temperature. Whenever possible, samples should be stored under water until required for testing. Samples should not be de-moulded until required for testing as unsealed samples will dry rapidly and even a few minutes' exposure on a laboratory bench may lead to significant water loss.

6.2. Laboratory testing

As noted in Section 6.1.1 cement-bentonite materials are strongly alkaline and may react with and damage any aluminium or other alkali-reactive material in the test equipment. It is therefore important to check that all test equipment etc. is inert to the slurry, its leachate and any special permeants, e.g. samples of the liquids to be retained by the wall.

6.2.1. Approved laboratory

Cement-bentonite materials are a distinct class of materials and their behaviour differs significantly from that of soils or concrete. Not all laboratories will be appropriately experienced to undertake the necessary testing. Laboratories having NAMAS accreditation of the relevant tests or equivalent are desirable.

6.2.2. Scheduling of tests

The requirement of Clause 6.1.4 in the Specification that samples shall not be transported until at least 14 days old will necessitate careful scheduling if tests at 28 days or younger are required.

It should be noted that this section also sets out the minimum testing requirements for the cut-off slurry unless other requirements are included in the Particular Specification. These are:

- two unconfined compressive strength tests, and
- two permeability tests

for 1000 m^2 of wall.

6.2.3. Preparation and care of samples

When preparing samples for testing in the laboratory, it may be necessary to trim damaged or irregular material from the ends of the samples. It follows that sample tubes should be longer than the required length of the test specimen, but see Section 6.1.1 on damage during extrusion.

As already noted, samples of cut-off wall slurries will be relatively fragile and should be treated as if they were sensitive clay. Before samples are extruded from tubes, the ends of the tubes should be checked to be free of burrs, adhering wax (see Section 6.1.1) etc., so that the samples can be extruded cleanly.

The requirement that there shall be no sub-sampling of smaller diameter samples from the original sample is discussed at Section 6.1.1 of these Notes for Guidance.

6.2.4. Tests

6.2.4.1. Triaxial permeability tests

Unless otherwise specified in the Particular Specification the test should be in accordance with BS 1377:1990, Part 6, Clause 6. This standard requires that samples are tested under a back pressure and that a specified minimum saturation is achieved before permeation is started. Cement-bentonite materials are relatively stiff and high confining pressures can be necessary to demonstrate the required saturation. For less stiff materials, this may be achieved at a lower absolute degree of saturation. Great care has to be taken when following the saturation procedure. If the back pressure is allowed to fall for any reason, substantial confining pressures, sufficient to cause damage, could be exerted on the sample.

The required saturation procedure will lead to an extended test time. It may be preferable to accept a lower saturation criterion, but require a longer time under permeation. Indeed some practitioners prefer to start permeation of the sample without the saturation stage and this appears to have little effect on the results, provided that the samples have been stored under water.

It is becoming common practice to require that proving trials are carried out on cut-off mixes whereby they are permeated with contaminated groundwaters or landfill leachates. Such tests are entirely reasonable but it must be appreciated that it may take a very substantial time for any deleterious effects be manifested. Short-term tests of a few days or months duration may give spurious and entirely erroneous indication of material performance. To demonstrate this point, consider, for example, the effects of permeation with an imaginary but particularly damaging fluid that completely destroys the impermeability of all material with which it has reacted. If this fluid was set to permeate through a sample of cut-off material by the time the reaction front had reached the middle, the first half of the sample would be of effectively infinite permeability. However, the remainder would still have its original permeability and thus the overall permeability would have increased by only a factor of 2. This could easily pass unnoticed or even have been reversed, if the second half of the sample had, in the meantime, undergone a modest time-dependent decrease in permeability. This could easily occur as the permeation time to react with and destroy half the sample is likely to be of the order of tens of years in a material of permeability 1×10^{-9} m/s. Indeed it may be greater than the design life of the cut-off wall, in which case the permeability of the reacted material may cease to be a design criterion and the permeability of the remaining unreacted material at any time will be the controlling factor.

It should be remembered that landfill leachate and other contaminated ground waters may not be chemically stable. If such fluids are used in permeation tests breakdown may occur and gas generation (e.g. methane) may disturb or reverse the permeant flow.

6.2.4.2. Unconfined compression strength tests

This test is to be carried out in accordance with BS 1377:1990 Part 7, Clause 7.

6.3. In-situ testing

On a few sites in the UK, test boxes have been installed to check the in-situ permeability of the wall. These have been formed by constructing a short length of wall parallel to the main wall and linking it to the main wall with cross walls. In general, the in-situ permeability has been found to be in reasonable agreement with the permeability measured in the laboratory, although it must be allowed that there are many uncertainties with a box test, and that it is difficult to prove a permeability of 1×10^{-9} m/s. The effective porosity of the soil in the box (and thus the water storage) may be uncertain and flow through the base of the box may be significant yet difficult to assess. If it is required to prove a wall to 1×10^{-9} m/s permeability, the test duration may have to be several months, and one or more pumps need to be installed within the box, and a constant water level maintained with measurement of flow. Alternative procedures which have been reported include:

- the insertion of pressure monitoring probes into the wall,
- the insertion of piezocones,
- precise analysis of pressure responses in boreholes adjacent to a wall following the injection of water at a controlled flow rate for a defined period.

An assessment of the various methods has been undertaken by the Building Research Establishment (Tedd *et al.*, 1995, 1997). Falling head permeability tests using the BRE packer system placed inside a small hole drilled in the middle of the cut-off wall have provided satisfactory measurements at a number of sites. However, problems have been experienced where quantities of the surrounding ground are present in the wall. Use of the piezocone has not provided representative permeability results. The subject of in-situ permeability assessment should be regarded as developing. The bibliography provides further information on these methods.

6.4. Geomembrane

6.4.1. Geomembrane conformance testing

The conformance testing is intended to check that the geomembrane has met the requirements of the Specification in terms of the NSF standard (see Section 3.2.1) and the additional requirements of Table 1. Further testing, as detailed in of the Particular Specification, may be required by the Engineer. However, some geomembrane tests take several weeks to carry out, and programming such testing should take into account the duration of the site works.

6.4.2. Geomembrane non-destructive testing

All welded joints should be subjected to non-destructive testing as a minimum. The contractor should propose the testing for the interlock system if requirements have not been included in the Particular Specification.

6.4.3. Geomembrane qualitative destructive testing

The requirement for qualitative destructive testing is based on the theory that the geomembrane interlock should not be a weak link in the barrier system. A possible procedure for qualitative strength testing is as follows:

A 25 mm wide field tab should be cut from the geomembrane-welded joint-interlock detail at the top or bottom of a panel. Each tab should include a minimum of 50 mm of parent geomembrane sheet in addition to the interlock detail and should be cut normal to the interlock. Pairs of tabs from adjacent panels can then be assembled and tested by stretching to failure with a hand-cranked tensiometer.

The interlock may be deemed to have passed qualitative destructive testing if the failure occurs solely in the parent material and does not enter either the weld or the interlock. The interlock may be deemed to have failed if the interlock comes apart or the weld fails prior to geomembrane sheet failure.

It is also important that the design of the interlocking system should not compromise the permeability of the overall geomembrane system. It is likely that interlocks will be more permeable than intact sheets of the membrane, and if this is a matter of concern it must be addressed in the Particular Specification by, for example, specifying an acceptable leakage rate and text procedure.

6.4.4. Geomembrane quantitative destructive testing

Very little information is currently available on quantitative destructive testing of geomembrane interlocks, and research work is ongoing. It is suggested that specification values should not be given in the Particular Specification, because at present no testing methodology or database of results is available. Work on the tensile strength of geomembrane panel interlock joints has been reported by Dixon *et al.* (1997).

7. Records

A national database of cut-off wall mix data has been established at the Building Research Establishment. This will enable wall specifications to be to be checked against actual performance, etc.

8. Post-construction monitoring

At the present time the design of cut-off walls is much constrained by the lack of long-term data on performance. Engineers and contractors ought to be made aware of the benefits that may accrue to them and the industry generally from such testing.

Also, owners and occupiers should be made aware that without post-construction verification and on-going monitoring of cut-off wall performance, land values may be compromised and sale/transfer may be much complicated.

Bibliography

References and further reading

Barker P. J. and Esnault A. (1992). Developments in techniques of ground treatment for the protection and rehabiltation of the environment. *Proc. of the 2nd Int. Conf. on Construction on Polluted and Marginal Land.* Brunel University, pp. 201–214. Engineering Technics Press.

Barker P. J., Esnault A. and Braithwaite P. (1997). Containment barrier at Pride Park, Derby, England. *International containment technology conference.* St Petersburg, Florida. pp. 95–103. Also *Land Contamination & Reclamation*, vol. 6, no. 1, 1997, pp. 217–223.

Burke G. K., Crockford R. M. and Achhorner F. N. (1997). Case histories portraying different methods of installing liners for vertical barriers. *International Containment Technology Conference.* St Petersburg, Florida. pp. 221–228

Chipp. P. N. (1990). Geotechnical Containment Measures for Pollution Control. *Proc. of the 1st Int. Conf. on Construction on Polluted and Marginal Land.* Brunel University, Engineering Technics Press. pp. 103–116.

Dixon N., Jones D. R. V. and Nicholas A. (1997). Tensile strength of geomembrane interlock joints. *Ground Engineering* vol. 30, no. 4. pp. 42–44.

Esnault A. (1994). Containment for polluted sites: new approach to choice of materials. *Proc. of the 3rd Int. Conf. on Construction on Polluted and Marginal Land.* Brunel University, Engineering Technics Press. pp. 147–151.

Garvin S. L., Tedd P. and Paul V. (1994). Research on the performance of bentonite-cement containment barriers in the United Kingdom. *2nd Int. Symp. on the Environmental Contamination in Central and Eastern Europe,* Budapest, pp. 414–416.

Garvin S. L. and Hayles C. S. (1999). The chemical compatibility of cement bentonite cut-off wall material. *Construction and Building Materials.* No. 13. pp. 329–341.

GLR (1993). *Geotechnics of Landfill Works Technical Recommendations— GLR.* Edited by the German Geotechnical Society for the International Society of Soil Mechanics and Foundation Engineering. Ernst & Sohn.

Jefferis S. A. (1981). Bentonite-cement slurries for hydraulic cut-offs. *Proc. 10th ICSMFE,* Stockholm. Balkema, Rotterdam. pp. 435–440.

Jefferis S. A. (1992). Contaminant-grout interaction. *ASCE Speciality Conference. Grouting, Soil improvement and Geosynthetics,* New Orleans, pp. 1393–1402.

Jefferis S. A. (1992). Slurries and Grouts. Chapter 48. *Construction materials reference book.* Ed. Doran, D K. Butterworths, 1992.

Jefferis S. A. (1996). Contaminant chemistry: friend or foe? Exploiting contaminant/barrier interaction. *Proc. Conf. Chemical Containment of Wastes in the Geosphere.* British Geological Survey, Nottingham.

Jefferis S. A. (1997). The origins of the slurry trench and a review of cement-bentonite cut-off walls in the UK. *International Containment Technology Conference,* St Petersburg, Florida. pp. 52–61.

Manassero M., Fratalocchi E., Pasqualini E., Spanna C. and Vega F.

(1995). Containment with vertical cutoff walls. *Proc. Geoenvironment 2000.* ASCE. pp. 1142–1172.

Manassero M. (1994). Hydraulic Conductivity Assessment of Slurry Wall Piczocone Tests. *Journal of Geotechnical Engineering.* ASCE vol. 120, no. 10, pp. 1725–1746.

Moura M. L., Weststrate F. A. and Loxham M. (1991). The effects of contaminants on the properties of cement-bentonite cut-off walls. *Environmental Pollution 1*—ICEP. pp. 322–325.

Privett K. D., Matthews S. C. and Hodges R. A. (1996). *'Barriers, Liners and Cover Systems for Containment and Control of Land Contamination',* CIRIA Special Publication 124.

Tedd P., Paul V. and Lomax C. (1995). Investigation of an eight year old slurry trench wall. Green '93. *Int. Symp. on Waste Disposal by Landfill.* Bolton Institute. Balkema 1995. pp. 581–590.

Tedd P., Holton I. R., Butcher A. P., Wallace S. and Daly P. J. (1997). Investigation of the performance of cement-bentonite cut-off wall in aggressive ground at a disused gasworks site. *International Containment Technology Conference.* St Petersburg, Florida. pp. 125–132. Also *Land Contamination & Reclamation*, vol. 5, no. 3, pp. 217–223.

Tedd P., Butcher A. P. and Powell J. J. M. (1997). Assessment of the piezocone to measure the in-situ properties of cement bentonite cut-off walls. *Proc. of Conf. on Contaminated ground:fate of pollutants and remediation.* Cardiff. Thomas Telford, pp. 48–55.

Tedd P., Quarterman R. S. T. and Holton I. R. (1995). Development of an instrument to measure in-situ permeability of slurry trench cut-off walls. *4th Int. Symp. on Field Measurements in Geomechanics.* Bergamo, Italy. pp. 441–446.

Xanthakos P. (1979). *Slurry Walls.* McGraw-Hill, New York.

Standards and geomembrane test methods

ASTM D638-95, *Test Methods for the Tensile Properties of Plastics.* Philadelphia.

ASTM D1004-94, *Test Method for Initial Tear Resistance of Plastic Film and Sheeting.* Philadelphia.

ASTM D1505A-90, *Test Method for Density of Plastics by the Density-Gradient Technique.* Philadelphia.

ASTM D1593-92, *Specification for non rigid Vinyl Chloride Plastic sheeting.* Philadelphia.

ASTM D1603-94, *Test Method for Carbon Black of Olefin Plastics.* Philadelphia.

ASTM D3015-95, *Practice for Microscopical Examination of Pigment Dispersion in Plastic Compounds.* Philadelphia.

ASTM D3895-95, *Test Method for Oxidation-Induction Time of Polyolefins by Differential Scanning Calorimetry.* Philadelphia.

ASTM D5397-95, *Test Method for evaluation of Stress Crack Resistance of Polyolefin Geomembranes Using Notched Constant Tensile Load Test.* Philadelphia.

BS 12:1991, *British Standard Specification for Portland cement.* British Standards Institution (BSI), London.

BS 1377:1990, *British Standard methods of test for soils for civil engineering purposes.* Parts 6, 7 and 8. BSI, London.

BS 6699:1992, *British Standard Specification for ground granulated blastfurnace slag for use in Portland Cement.* BSI, London.

BS 3892 Part 1:1993, *British Standard Specification for pulverised fuel ash for use as a cementitious component in structural concrete.* BSI, London.

BSEN 1538:1999, *Execution of Special Geotechnical Work—Diaphragm Walls.* BSI, London.

EPA/530/SW-09/069 (1989). *The fabrication of polyethylene FML field seams*. US Environmental Protection Agency, Washington, DC.

FTMS 101C Method 2065, *Puncture resistance and elongation test (1/8 inch radius probe method)*, Federal Test Method Standard, March 1980.

National Sanitary Foundation. *Flexible Membrane Liners*, Standard Number 54, 1993.

Oil Companies Materials Association (OCMA) (1973). Specification No. DFCP–4, *Drilling Fluid Materials Bentonite.* Also Engineering Equipment and Materials Users' Association Publication No 163, *Drilling Fluid Materials Bentonite.*

The Institution of Civil Engineers
Construction Industry Research
and Information Association
Building Research Establishment

Specification for the construction of slurry trench cut-off walls
as barriers to pollution migration

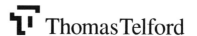

Published by Thomas Telford Publishing, Thomas Telford Limited, 1 Heron Quay, London E14 4JD

URL www.t-telford.co.uk

First published 1999

Distributors for Thomas Telford books are
USA: ASCE Press, 1801 Alexander Bell Drive, Reston, VA 20191-4400, USA
Japan: Maruzen Co. Ltd, Book Department, 3–10 Nihonbashi 2-chome, Chuo-ku, Tokyo 103
Australia: DA Books and Journals, 648 Whitehorse Road, Mitcham 3132, Victoria

Also available from Thomas Telford

ICE Specifications
- Specification for ground investigation 0 7277 1984 X
- Model specification for tunnelling 0 7277 2588 2
- Specification for piling and embedded retaining walls 0 7277 2566 1
- Essential guide to the Specification for piling and embedded retaining walls 0 7277 2738 1
- Specification for ground treatment 0 7277 0388 9

Environmental engineering
- Advanced landfill liner systems 0 7277 2590 4
- Contaminated land and its reclamation 0 7277 2595 5
- Environmental assessment 0 7277 2612 9
- Environmental law for the construction industry 0 7277 2611 0
- Final covers for solid waste landfills and abandoned dumps 0 7277 2643 9
- Geoenvironmental engineering: contaminated ground – fate of pollutants and remediation (1996) 0 7277 2840 7; management and remediation (1999) 0 7277 2606 4

Cover shows construction of a 30 m deep slurry trench cut-off wall (courtesy of Keller Ground Engineering).

A catalogue record for this book is available from the British Library

ISBN: 0 7277 2625 0

© The Institution of Civil Engineers, the Construction Industry Research and Information Association and the Building Research Establishment, 1999

All rights, including translation reserved. Except for fair copying, no part of this publication may be reproduced, stored in a retrieval system or transmitted in any form or by any means, electronic, mechanical, photocopying or otherwise, without the prior written permission of the Books Publisher, Thomas Telford Publishing, Thomas Telford Limited, 1 Heron Quay, London E14 4JD.

This book is published on the understanding that its publication does not necessarily imply that statements and/or opinions expressed in it are or reflect the views or opinions of the publishers.

Whilst every effort has been made to ensure this document constitutes an accurate statement and guide, neither the authors, the Institution of Civil Engineers, the Building Research Establishment Limited, nor the Construction Industry Research and Information Association can accept any liability for any loss or damage which may be suffered by any person as a result of the use in any way of the information contained herein.

Typeset in Great Britain by Apek Digital Limited, Bristol.

Printed and bound in Great Britain by Selwood Printing Limited, West Sussex.